What's Science Ever Done for Us?

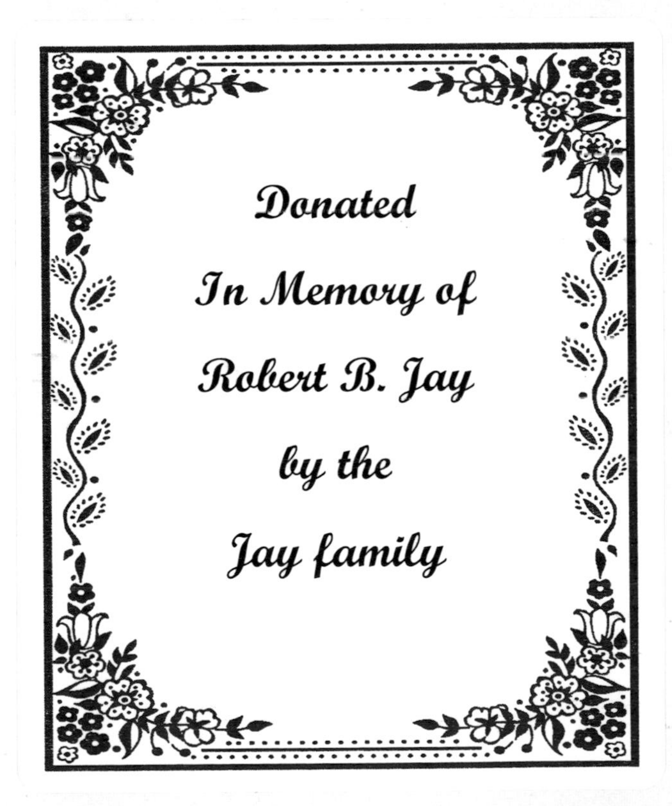

What's Science Ever Done for Us?

What *The Simpsons* Can Teach Us about Physics, Robots, Life, and the Universe

PAUL HALPERN

John Wiley & Sons, Inc.

Published by John Wiley & Sons, Inc., Hoboken, New Jersey
Published simultaneously in Canada

Wiley Bicentennial Logo: Richard J. Pacifico

Design and composition by Navta Associates, Inc.

For general information about our other products and services, please contact our Customer Care Department within the United States at (800) 762-2974, outside the United States at (317) 572-3993 or fax (317) 572-4002.

Wiley also publishes its books in a variety of electronic formats. Some content that appears in print may not be available in electronic books. For more information about Wiley products, visit our web site at www.wiley.com.

Library of Congress Cataloging-in-Publication Data:

Halpern, Paul, date.
What's science ever done for us? : what the Simpsons can teach us about physics, robots, life, and the universe / Paul Halpern.
p. cm.
Includes bibliographical references and index.
ISBN 978-0-470-11460-5 (pbk.)
1. Science—Popular works. 2. Technology—Popular works. 3. Simpsons (Television program) I. Title.
Q162.H3154 2007
500—dc22

2007002542

Printed in the United States of America

10 9 8 7 6 5 4 3 2 1

For my sons, Eli and Aden

"Science? What's science ever done for us?"

—*Moe Szyslak, bartender, "Lisa the Skeptic"*

CONTENTS

INTRODUCTION

Learning Science from Springfield's Nuclear Family

Ah, there's nothing more exciting than science. You get all the fun of sitting still, being quiet, writing down numbers, paying attention. Science has it all.

—Principal Seymour Skinner, "Bart's Comet"

Hurray for science! Woo!

—Bart Simpson, "Bart's Comet"

The cumulus clouds gather and part, revealing the endless blue skies over the town of Springfield. All seems sunny and bright, from the shiny rows of houses to the gleaming stores and taverns. Towering above them all are the friendly cooling towers of Springfield's expertly run nuclear plant—the very model of efficiency, at least according to its paperwork. Residents benefit from the warmth and sustenance provided by this central hearth, a steady source of energy and jobs.

If you live in Springfield—or any other town, for that matter—you cannot help but be affected by science. If your home isn't lit by nuclear power, then it's fueled by coal, kerosene, wind power, hydroelectric energy, solar power, or another means. Even if you live in a tent on the beach, there's the sun, the moon, and stars—and perhaps a roaring campfire—bringing you light and heat. For those who reside in caves deep underground, there are glowworms. Each source of power runs through a unique physical mechanism. You simply cannot escape science.

The benefactor behind Springfield's veritable utopia—the paternal figure from whom the precious milk of power flows—is none other than Springfield's leading entrepreneur, C. Montgomery Burns. He doesn't mind if people are kept in the dark—about science, that is. As long as their pennies for each ticking kilowatt-hour flow into his coffers, he's quite elated. "Exx-cellent," he often cackles to his loyal assistant, Wayland Smithers.

Keeping the plant and the town out of danger is someone who *ought* to know a lot about science, America's everyman, Homer Jay Simpson. By occupation, if not by experience, he's well linked to science—some have even speculated that he's Darwin's missing link. His job as plant safety inspector requires the highest technological know-how—determining for which warning messages he needs to press the buttons on his monitor and which offer him time to take a donut break or a nap. Although not a classic intellectual, Homer demonstrates his true pensiveness when faced with any challenging issue. Ask him even the most difficult question and you can count on his response. You can almost see the wheels turning—behind him on the machinery as he stares off into space. Disinclined to speak too soon, he pauses for a while, then hesitates. After a long meaningful silence, as if he were in an Ingmar Bergman movie, he pauses again. He hesitates once more, lest the wrong words roll off of his tongue. Zzzzzz. Sometimes even the most pressing problems have a way of resolving themselves.

When it is time for lunch at the plant, Homer shares lighthearted moments with his pals Lenny Leonard and Carl Carlson. Although Carl has a master's degree in nuclear physics, he and

Lenny are just regular beer-drinking guys. Lenny has a chronic eye problem, so he makes sure to aim his drinking glass properly. Lenny and Carl also often join Homer after work at a tavern run by the cynical and sometimes suicidal Moe Szyslak. Moe is not exactly fond of science; he once dissed its value shortly before using a voice-activated TV (see the title quote of this book). Running a tavern ain't rocket science, so he never bothered to learn that field.

Springfield, in a nutshell, is full of stark contrasts in its attitude toward science. Having a nuclear power plant in the heart of town that provides the bulk of its jobs forces the inhabitants to confront technological issues on a daily basis. Moreover, the town is strangely faced with more than its lion's share of calamities—from colliding comets and invading aliens to black holes materializing in home supply stores and the sun overhead being blotted out—the last being a fiendish plot hatched by Burns. You would think that the townspeople would be crying out for solid scientific know-how. Yet what expertise exists is often downplayed or ignored. The town's resident genius, John Frink, a bona fide nutty professor (as in the Jerry Lewis original film, not the sequel), is treated like a virtual pariah. Perhaps it's his lack of social grace and incoherent way of speaking—with ample use of nonsense words such as *glaven*—that isolate him from his would-be peers. Nevertheless, given his extraordinary inventiveness, you'd think they'd reach out to him—maybe even elect him mayor instead of the pandering, philandering Joe Quimby, who presently serves in that office.

In medicine, too, mediocrity often trumps expertise. Though the town has a perfectly capable physician, Dr. Julius Hibbert, patients often turn to the quackery of Dr. Nick Riviera instead. Maybe that's because Dr. Hibbert charges a fortune and chuckles during inopportune moments such as delivering devastating diagnoses, or even making one up as a joke. Comforting bedside manner, he realizes, isn't covered by most insurance plans. Dr. Nick, on the other hand, has the medical expertise of a tree stump, but he's superficially friendly, doesn't laugh when you ask him to do wacky procedures, and is relatively cheap.

Many Springfield residents attend the church of Reverend

Timothy Lovejoy, who seems downright hostile to science. Among the most devout of Lovejoy's flock is Homer's affable, straitlaced neighbor Ned Flanders. Homer often cringes when Flanders calls out "Hi-dily-ho neighborino" and other variations on this greeting, bracing himself for a stern moral critique. "Stupid Flanders," as Homer calls him, doesn't seem to know how to kick back and enjoy life—at least from the perspective of a television-addicted, donut-munching beer-guzzler. Yet Flanders usually seems joyous in his faith, finding simple pleasure in helping the downtrodden. It is when faith and science tell different tales that Flanders's anxiety piques and he primes himself for battle, usually with Lovejoy's support. For example, together they have fought to eliminate all mention of evolution from Springfield's textbooks.

Where does the principal of Springfield's elementary school, Seymour Skinner, stand on this? He clearly loves science, as demonstrated by his amateur astronomical pursuits in which he hopes to find and name his own comet. He found one once but was scooped by a certain Principal Kohoutek. Yet, with the backbone of a jellyfish, Skinner often loses control over the school's curriculum. From his mother, Agnes, down to his pupils—and even his erstwhile girlfriend/fiancée, teacher Edna Krabappel—no one seems to respect him. District superintendent Chalmers constantly bawls him out, leaving him precious little wiggle room. He has only custodian Groundskeeper Willie, a proud Scotsman who does undignified menial chores, to kick around. Unless, that is, he has been temporarily demoted to Willie's assistant, as when he is briefly replaced as principal due to inappropriate comments about girls and math.

Other characters on the show are too caught up in their hobbies to spend much time worrying about science. School-bus driver Otto Mann's only connections with chemistry are the substances he ingests and heavy metal music. Comedian Krusty the Clown, born Herschel Krustofski, is too busy preparing his laugh-riot television program, running his fast-food empire, and trying to reconcile with his rabbi father. Krusty's former assistant, Robert "Sideshow Bob" Terwilliger III, is obsessed with murdering a certain young tyke he despises. Fellow criminal Snake Jailbird is determined to earn a fortune through armed robbery. His main target, Kwik-E-Mart convenience store manager Apu Nahasapeemapetilon, can only find time, between holdups, to sell flavorful "Squishees" and protect his magazine rack, which is emphatically not a lending library, from perusal. That's a shame, because he has a Ph.D. in computer science that has gone to little use except to try to impress women during his bachelor days. Another shopkeeper, Jeff Albertson, better known as the "Comic Book Guy," at least has a passion for science fiction. In his store, the Android's Dungeon and Baseball Card Shop, he sells more informative magazines such as the illustrated adventures of the famed Radioactive Man with his sidekick, Fallout Boy, than can be found in mere mini-marts.

The prospects for true science flourishing in Springfield would seem nearly hopeless if it weren't for several of its most illustrious (but rarely seen) residents. The late acclaimed paleontologist Stephen Jay Gould lives on in one of the finest episodes of the series when he appears as himself working in the Museum of Natural History. Gould evaluates strange skeletal remains found beneath a building site. Another famous scientist, the Cambridge physicist Stephen Hawking, pops up in two episodes. Reportedly, Hawking is a great fan of the show and was "vastly proud of his appearance."[1] He seemed to have a lot of fun with his roles—especially his second appearance, in which he works at the local Little Caesars pizzeria. Unlike with Frink, the townspeople appear to have more respect for Hawking's opinions; it's a shame that he isn't around more often to correct their misconceptions. A third highly accomplished scientist who has appeared is Dudley Herschbach, the co-recipient of the 1986 Nobel Prize in Chemistry, whose brief role in one episode involves awarding a Nobel Prize to Frink.

Yet another notable who has made two "appearances" on the Simpsons is the reclusive author Thomas Pynchon; his character is shown each time with a paper bag on his head. Though not a scientist, Pynchon studied engineering physics for two years at Cornell. Many of his writings contain ample allusions to science, from "Entropy," one of his first short stories, to his renowned novel *Gravity's Rainbow*, and finally to his recent novel *Against the Day*, which includes physicist Nikola Tesla as a character. To the great surprise and pleasure of his fans, although Pynchon has declined all interviews, photographs, and recordings for decades, he premiered his voice and verbal wit on the show.

Any town listing Gould, Hawking, Herschbach, and Pynchon as residents (or at least visitors) would seem to have great potential for a healthy attitude toward science, particularly if the younger generation could be persuaded to follow in these illustrious thinkers' footsteps. Could it be that the indifference or hostility toward science expressed by certain Springfield grownups could be overcome by the savvy of youth? There the hope lies in an extraordinary young scholar, Homer's precocious eight-year-old daughter, Lisa.

Intellectually, Lisa towers above the fellow students in her school, save perhaps brainy fourth-grader Martin Prince. Whenever Principal Skinner wants to impress visitors with a "typical student" who demonstrates the school's high caliber, Lisa is showcased. Other pupils range from babyish, clueless Ralph Wiggum—whose father, Clancy, is the police chief—and Lisa's awkward, bespectacled wooer, Milhouse Van Houten, to the school bullies who love to beat up such helpless kids: Jimbo Jones, Dolph, Kearney, and their juvenile-delinquent leader, Nelson Muntz. Nelson's catchphrase "Ha ha!" repeated every time he witnesses a misfortune or foible, is no match for Lisa's soft-spoken eloquence. Similarly, other schoolmates, from the twins Sherri and Terri to German exchange student Üter, offer no real competition.

In Lisa's family too, though she is the second smallest, she is clearly the intellectual giant. Despite Homer's technological job and active imagination—as expressed in off-the-wall daydreams—he is one crayon short of a full pack. In fact the missing crayon is lodged in his brain, as revealed in the episode "Homr," loosely based on the classic story "Flowers for Algernon." When the crayon is surgically removed, Homer's IQ goes up by 50 points. Heightened intellect, though, has its drawbacks. Homer, realizing all of the safety violations at his plant, reports it to the Nuclear Regulatory Commission, resulting in its temporary closing. Lenny and Carl, now out of a job, are resentful, to say the least. Finding that intelligence can't buy happiness, Homer asks Moe, who performs surgical procedures on the sly, to insert the crayon back into his brain. Since then, Homer seems even more dim-witted, if that is possible. Despite Homer's obvious faults, Lisa loves her dad with all her heart.

Marge, neé Marjorie Bouvier, Homer's wife and the matron of the family, appears to be the next-brightest of the bunch (of at least the *speaking* members of the family), if only for her outstanding common sense and many practical talents, including an aptitude for mechanics. In high school she enjoyed calculus until Homer convinced her to give it up. Considering her substantial abilities, she could certainly be more assertive. She's accepting to a fault, often refusing to take sides for fear of offending someone. Her unwillingness to commit herself often exasperates Lisa, who would like her mother to weigh the evidence and take a stance. Nonetheless, it often turns out that Lisa has conflicting opinions herself about facts versus faith that she is afraid to express lest she seem less than a true scientific thinker. During those moments of doubt, she can better understand her mother's balanced views.

Maggie, the baby in the family, is necessarily a big question mark, since we have never really heard her express herself—just some babbling noises, a few first words (like "Daddy"), and mainly the sucking sounds of her ever-present pacifier. Even in episodes speculating about the family's future, she still doesn't have a chance to say anything. Only in some of the annual "Treehouse of Horror" Halloween episodes—considered nightmares, stories, or alternative realities, not part of the real family history—does Maggie speak in full sentences. Thus she could well turn out to be the smartest Simpson, a point hinted at in a number of episodes. For example, during a family Scrabble game she happens to spell out "EMCSQU" ($E = mc^2$) with her building blocks.

Finally, we come to the enfant terrible of the show, the ten-year-old boy who turned "Eat my shorts!" and "Don't have a cow!" into international catchphrases, immortalized on T-shirts, in comic books, and the like. He's the skateboarding kid whose smash-hit song "Do the Bartman!" single-handedly rescued commercial radio from utter oblivion. (Admittedly, I exaggerate here, but it's a fun novelty tune.) I speak, of course, of none other than Bartholomew Simpson, better known as Bart—or as Homer calls him while wringing his neck, "Why you little . . . !"

Although Bart has a keen curiosity, he finds school an utter challenge and is much happier pulling pranks. When it comes to scientific discovery, he tends to be more of a passive observer—stumbling accidentally onto novel findings—than an original thinker in his own right. For instance, when Skinner punishes Bart by forcing him to engage in astronomy, Bart ends up spotting his own comet. He is happy and capable when playing a video game that has scientific content, until he realizes that it is educational and backs off. He'll experiment with mixing chemicals together, as long as it's to make a cool-looking explosion rather than for an actual assignment. With resounding antipathy toward formal learning, he can nevertheless easily be tricked into gaining knowledge.

Could someone like Bart learn science from an informal source, such as a comic book or a cartoon? Without a doubt. If Radioactive Man, his favorite comic book series, or *The Itchy and Scratchy Show*, his beloved television cartoon show, urged aficionados to perform certain chemistry or physics projects to help out the characters, and even to investigate the history and background of these experiments, you bet he would rise to the task. Many kids quickly learn the difference between "fun science" and what they—gasp!—are graded on. Naturally they tend to gravitate toward the former, except perhaps to cram information before a test.

In that regard, *The Simpsons* offers a perfect venue for informal science education. It's one of the few comedy programs with no laugh track—and plenty of brains. In the absence of an authority telling you when to laugh or learn, you are forced to sift through cutting sarcasm, conflicting opinions, and occasionally even sly misrepresentations to figure out the truth.

A number of writers on the show have scientific connections and love to refer to their subjects. These include David X. Cohen, who has a bachelor's degree in physics from Harvard and a master's in computer science from U.C. Berkeley; Ken Keeler, who has a

Ph.D. in applied math from Harvard; Bill Odenkirk, who has a Ph.D. in inorganic chemistry from the University of Chicago; and Al Jean, the executive producer and head writer, who has a bachelor's degree in mathematics from Harvard. Another writer, Jeff Westbrook, has a Ph.D. in computer science from Princeton and was an associate professor of computer science at Yale several years before he joined the series. He was involved with the 2006 episode "Girls Just Want to Have Sums," related to the recent controversy at Harvard concerning comments made by its president about women in mathematics.[2]

Given the expert background of the show's stable of writers, it's not surprising that ample doses of science, math, and technology, offering tastes of many different fields, are sprinkled throughout many of the episodes. Topics include everything from astronomy to zoology and genetics to robotics; you just have to dig deep sometimes to uncover the facts. Like Kent Brockman, the TV news anchor on the show, you need to be an investigative reporter—that's part of the fun of scientific discovery. Instead of revealing the gossip behind staid celebrity veneers, you'll be uncovering the true scientific facts behind the show's contagious silliness. As Krusty might say in one of his reflective moods, there's often a serious story behind the laughter. Hey! Hey!

Academics have already stumbled upon the show's serious undercurrent. It is rare for a cartoon on television to trigger intellectual discussion and even generate published articles. Yet *The Simpsons* has inspired publications about health care, psychology, evolution, and other issues. It is a series watched by many scientists and therefore scrutinized for its accuracy and implications in an unprecedented way. Each yuk, har-har, and guffaw has been laboratory tested for quality, kids, so pay close attention!

In that vein, this book is meant to be a field guide to the science behind the series, so—even while you are rolling on the floor in hysterics—you can appreciate and learn from its abundant references to biology, physics, astronomy, mathematics, and other fields. Impress your friends and baffle your enemies with your detailed knowledge of the background behind the episodes. Satisfy your intellectual curiosity while warming up your house with the radiance from your television set. Quench your burning questions with the invigorating Buzz Cola of scientific fact, available through the vending machine of the airwaves. Just gather on your couch and let the lessons begin.

Throughout its run of more than two decades (including several seasons as part of *The Tracey Ullman Show*), various segments of *The Simpsons* have raised many intriguing issues about the workings of contemporary science. The breadth of these questions is astonishing. For example, how do paleontologists such as Gould determine the age of skeletal remains, such as those Lisa discovers and brings to him? What factors cause mutations, such as the one that spawns Blinky the Three-Eyed Fish, which swims in Springfield's polluted waters? Why can't the stars and planets over Springfield be seen clearly at night? Could androids, such as the robot that replaced Bart in one of the Halloween episodes, ever have consciousness? Do toilets in the Northern and Southern Hemispheres swirl in opposite directions, as Lisa purports in the episode where the family travels to Australia? What are comets made of, such as the one Bart discovers, and how could they threaten Earth? If there are extraterrestrials in space, why haven't they visited Earth or even contacted us, in the manner of Kang and Kodos, the resident aliens on the show? Can time be reversed or stopped, as Homer and Bart have done in various segments?

Before tackling these wide-ranging scientific issues, let's consider one of the deep mysteries of the series. It's related to what I call the "Marilyn Munster conundrum" concerning unusual diversity among family members (Marilyn was the only attractive, nonmonstrous Munster on the television show of that name) and is connected to ongoing debates about nature versus nurture. If Lisa is a Simpson, why is she so smart?

PART ONE

It's Alive!

I'm afraid you're stuck with your genes.

—Dr. Julius Hibbert, "Lisa the Simpson"

[T]here's nothing wrong with the Simpson genes.

—Homer Simpson, "Lisa the Simpson"

1

The Simpson Gene

Mundane families are all alike; every unusual family is unusual in its own way. The Simpsons are emphatically a breed unto themselves. Begin with Homer's fanatical cravings, bizarre non sequiturs, off-the-wall daydreams, childish single-minded pursuits, and overall obliviousness. Add to the lunacy Grandpa's bizarre, rambling stories, full of implausible, inconsistent recollections of World War II, and his wholly unexplained antipathy toward the state of Missouri. Mix in Bart's propensity for utter mischief and absolute disregard for authority. Watch them insult, scream at, and even try to strangle one another. Not even Tolstoy, who wrote much about dysfunctional families, could keep up with all the twists and turns of the crazy plot machinations, let alone of Bart's poor neck.

You can place the blame squarely on the male Simpsons. Amid the tempestuous cauldron that they lovingly call home, the female members of the family usually manage to keep their wits about them. Immersed in situations that would rattle even the steeliest nerves, they typically offer the calm voice of reason. Even the continuous chomp-chomp of Maggie's pacifier offers a sedate mantra that seems to put matters in perspective.

What could explain the profound differences between the male and the female Simpsons? Is it purely a matter of differing expectations and environmental conditions—in Bart's case, for example, a reduced supply of oxygen through his trachea that occurs at regular intervals—or could there be a genetic component? In the episode "Lisa the Simpson," this question comes to the fore when Lisa wonders if simply being in her family dooms her to daftness and finds considerable relief when she learns that her gender could spare her.

The episode starts off with Lisa fearing that she is losing her intellectual gifts, such as solving math problems and belting out jazz pieces on her saxophone. Lisa prides herself on her intellect—demonstrated, for instance, in another episode where she attends a Halloween costume contest dressed as Albert Einstein. She clearly doesn't want to grow up and be just like the rest of her family. Homer and Bart often embarrass her with their childish antics, Marge is not fully realized, and Lisa sincerely hopes that her sharp mind will propel her to better things. But what if her powers of thought sputter before they convey her to her rightful position in life, and she ends up just like the other family members?

Lisa's anxieties skyrocket when Grandpa tells her about the "Simpson gene," a genetic predisposition to mental decline that kicks in during mid-childhood. As young children, Grandpa explains, Simpsons act perfectly normal. Slowly, however, the Simpson gene triggers deterioration of the brain, leading to lives of utter mediocrity or worse. Naturally, Lisa is petrified that the same thing will happen to her.

In an attempt to dispel Grandpa's theory and cheer Lisa up,

Homer invites a number of their relatives over. He asks them to describe what they do for a living, hoping the reports will impress her. Some of the male Simpsons speak first and, to Lisa's horror, turn out to be failures. Her great-uncle Chet has bungled running a shrimp firm. Her second cousin Stanley just hangs out at the airport and shoots birds. Another runs in front of cars to collect insurance money. None of these men give Lisa much hope.

Fortunately, several Simpson women chime in with glowing accounts of successful careers. One of them, the highly articulate Dr. Simpson, explain that the faulty Simpson gene is carried on the Y chromosome and passed down only from male to male. Lisa realizes that it's just the Simpson men who are doomed; the women are all fine.

Not only does this revelation mean that Lisa will grow up normal, it also implies that any children she has would be safe, too. But for Bart and other male family members having kids would be risky. This genetic roulette is the exact opposite of baseball—if you strike out, you get a Homer.

It is an interesting theory, but could a single gene create such an intellectual disparity between women and men in a family? Intelligence is a complex issue, with smartness and success due to a variety of factors, environmental as well as genetic, many of which are not fully understood. Indeed such complexity is borne out in other episodes of the series, in which the male/female differences among Simpson family members are not so clear-cut. For example, in the episode "Oh Brother, Where Art Thou?" Homer meets his long-lost half-brother, Herb, who turns out to be wealthy and extremely successful. In "The Regina Monologues," Homer travels to England and encounters his long-lost half-sister, Abbie, who appears remarkably similar to him in voice, appearance, and seeming brightness. So on the face of it, Homer-like characteristics couldn't be solely a male thing; there must be other factors.

Moreover, as mentioned in the introduction, at least part of Homer's difficulties stems from a crayon wedged in his brain since he was a kid. Childhood traumas can in some cases cause impairment that extends into adulthood. Even without a specific incident, an overall environment hostile to learning could have profoundly negative repercussions throughout someone's life.

Children have an extraordinary capacity to adapt to whatever environment they are born into. The child who thrives in a nurturing, stimulating household might have faltered if born into a dreary, uncaring situation instead. Modeling themselves on their family members and friends, children often take on the attitudes and cultural norms of those around them. If a society radically changes its values—for example, renouncing violence after an age of militarism or becoming open and democratic after an era of totalitarianism—it's remarkable how quickly the bulk of its youth start to echo the new views. Thus environment and culture play tremendous roles in shaping the patterns of life.

Because of the profound influence of environmental factors, it's tempting to think that every child has unlimited potential to succeed in any area. Yet we must also recognize a genetic heritage that influences the pace of human development and the ultimate physical and mental limitations of individuals. No typical ten-year-old, no matter how extensively trained, could develop the strength to lift 400-pound weights or memorize all the names in the Chicago phone book. It would be foolish to expect that any kid practicing an instrument for ten hours each day could mimic the feats of Mozart or even develop enough proficiency to join a professional orchestra. Potential Olympic athletes must be identified at a very young age, not just for their abilities at the time, but also for their likely inherent potential.

The body's genome, or full set of genes, constitutes the codebook for how the body develops and functions. Each gene encodes a particular protein that typically serves a biological role, from the collagen in the skin to the muscle fibers in the heart. The two copies each of approximately 33,000 genes in the human body are arranged along 23 pairs of chromosomes. One copy of each gene comes from the mother and the other comes from the father, guaranteeing that everyone has a mixture of parental attributes.

Genes come in different sequence variations, called alleles. Each allele creates a difference in the constitution of the protein that a given gene encodes. For example, different alleles for the genes linked with eye color correspond to distinctive pigment proteins that in tandem could lead to variations in this trait. The specific pattern of genes is called the genotype. This should be distinguished from the phenotype, specifically how that pattern manifests itself as actual physical traits. Many different genetic patterns could end up producing the same trait—meaning that a range of genotypes could lead to the same phenotype. While phenotypes are often observed qualities, such as hair texture or whether or not someone can curl their tongue, determining a genotype generally requires genetic sequencing (mapping out the pattern of genes).

If chromosomes are the chapters of the body's encoding, and genes are the instructional pages with recipes for each protein, the specific sequence of bases on the helical, double-stranded molecules called deoxyribonucleic acid (DNA) constitutes the detailed language for these instructions. There are four different "letters" in the DNA "alphabet": the bases adenine, thymine, cytosine, and guanine, known as A, T, C, and G. Each base links with a partner on the opposite strand of the DNA: A with T and C with G. The particular arrangement of these bases produces the directions for manufacturing a multitude of different proteins.

Genes cannot synthesize proteins directly, however. Through a process called transcription, the coiled double helix of DNA creates single-stranded molecules called ribonucleic acid (RNA), which carry similar information but serve a different purpose. RNA differs from DNA in several ways, including its number of strands and the presence of the base uracil instead of thymine. One type of RNA, called messenger RNA (mRNA), forms a kind of protein assembly plant. Each set of three bases, called a codon, produces a specific type of amino acid. The particular chain of amino acids created in this process yields a certain type of protein.

Because cells in the human body carry (except for errors) identical versions of DNA, that cannot be the full story. When embryos develop in the womb, cells divide and differentiate, expressing their genetic content in different ways. Consequently, soon after conception, after enough divisions have taken place, cells start to specialize into skin cells, nerve cells, muscle cells, and so forth. The cell's relative position in the developing embryo seems to play an important part in this. The process of differentiation has long been one of the greatest mysteries in biology and is currently a vital topic of study.

The governing factor in heredity is the fact that chromosomes come in pairs—one set of genetic contributions from each parent. A given gene could appear in the form of either different or similar alleles, meaning there could be one or two copies of each allele. Alleles may be either dominant or recessive, depending on their biochemical properties. If an allele is dominant, then even if there is only one copy, the trait associated with that allele expresses itself and becomes part of the phenotype. For a recessive allele, in contrast, two copies must be present for that trait to appear. These rules were discovered by the nineteenth-century Czech botanist Gregor Mendel, who performed extensive studies of pea plant characteristics. He found, for instance, that tall alleles always dominated over short ones, meaning that tall plants bred with either tall or short plants always produced tall offspring.

Some inherited characteristics are specific to sex and manifest themselves differently for female and male offspring. The twenty-third pair of chromosomes, known as the sex chromosomes, is composed of two varieties, X and Y. Women almost always have an XX pair and men almost always have XY. (There are some rare conditions with other combinations.) The X chromosome is far bigger and has many more genes than the Y. With approximately 1,100 genes, consisting of more than 150 million base pairs, the X chromosome constitutes about 5 percent of the total number of human genes. Contrast that with the Y chromosome, which has only 78 genes. In recent years, these genes have been fully mapped out by the researchers Richard Wilson and David Page of Washington University in St. Louis. Wilson and Page noted that the genes in the

Y chromosome are mainly centered around the functioning of male reproduction—formation of the testes, sperm production, and so forth. Because these few genes are so important for the propagation of the species, the Y chromosome has evolved with multiple backup copies of the set. This duplication provides an assurance that even if one group of male reproduction genes is faulty, another set could express itself instead.

Hence, at least in terms of the Y chromosome, redundancy seems to be a critical male trait. That is to say that repetition, in genetics, is an important aspect of maleness. In other words, men, at least with regard to the Y chromosome's genes, often repeat themselves. Or how else could I put this. . . ?

Now that the genetic profile of the Y chromosome is well known, it does not appear to contain any gene that directly affects intelligence and common sense (unless you count teenage distractions due to the hormones of puberty). Thus the Simpson gene couldn't be found on the Y chromosome, and it couldn't be linked only to men. Alas, if such a gene existed, it could not be passed down exclusively from male to male, and therefore Lisa would have no firm guarantees of escaping its effect.

It's possible instead that such a gene could be on the X chromosome, a situation called sex-linked. Ironically, a sex-linked trait, though associated with an X chromosome gene, would appear more often in men if the causal allele happens to be recessive. That's because for women there's a choice between alleles from two different X chromosomes, but for men there's only one possibility. Hence, recessive alleles on a male's X chromosome are generally expressed.

A son receives his X chromosome exclusively from his mother. Therefore, if he inherits a sex-linked trait, it must have stemmed from the maternal side. Any sex-linked traits Bart has acquired, for example, would have been from Marge's genetic contribution, not Homer's. Similarly, Homer's male-pattern baldness, a sex-linked trait, could be chalked up to a recessive gene passed down from his mother, Mona, rather than his father, Abe.

There is one known sex-linked trait connected with aspects of intelligence—a hereditary condition called fragile X syndrome, so named because of a noticeable gap or fragile region in the X chromosome. This syndrome is due to changes to a gene called FMR1 that preclude it from producing a protein called FMRP (fragile X mental retardation protein). A particular triple sequence of bases on the FMR1 gene—cytosine-guanine-guanine (CGG)—is normally repeated about thirty times. For some individuals, an alteration occurs called a premutation that significantly increases the number of repetitions of that triplet up to two hundred times. Some researchers believe premutation of FMR1 could lead to subtle deficits in the intellectual or behavioral areas. If someone within a premutated version of FMR1 has children, her offspring have an increased chance of acquiring that gene in the fully mutated form. In that version, the CGG sequence is repeated more than two hundred times, usually triggering a process that prevents the production of FMRP and leads to fragile X syndrome. Fragile X syndrome has been associated with a number of effects, including cognitive and learning disabilities as well as alterations in physical appearance that emerge during adulthood. Aside from Down syndrome, an unrelated chromosomal disorder, scientists believe that fragile X syndrome is the leading genetic cause of mental impairment. Because it is sex-linked, fragile X syndrome affects many more men than women.

Not all inherited characteristics that affect males and females differently are sex-linked traits. Sometimes genes located on autosomes (nonsex chromosomes) respond differently to male and female biochemistry and produce distinct traits. For such a situation, these traits are called sex-influenced. Hence it is possible that a Simpson gene could be sex-influenced, rather than sex-linked. In that case, both Bart and Lisa could have inherited it from Homer, yet perhaps their dissimilar biochemistries caused it to respond in different ways.

Intelligence represents a very complex set of abilities that differs greatly from individual to individual. Because of varying definitions, researchers don't agree even on all the components of intelligence, let alone exactly which genes control it. It is also unclear how much depends on nature or nurture. Certain conditions that bear upon cognitive abilities, such as fragile X, have been mapped out, yet genetic research has a long way to go before being able to explain why family members, such as the Simpsons, act in such divergent ways.

Life has many mysteries, and the precise set of factors influencing Homer's erratic behavior appears to be one of them. He is a riddle wrapped in a mystery packed into stretchy blue trousers. Even the Human Genome Project could not unravel why Homer sapiens (as perhaps he could be classified) often operates with such bizarre motivations. How could we explain, for instance, why Homer would attempt to market a radiation-produced hybrid of tomatoes and tobacco?

2

You Say Tomato, I Say Tomacco

Some concepts need time to ripen, until they burst forth with delicious results. Other notions simply rot on the vine. It's hard to say where the idea of combining tomatoes with tobacco fits in—is it a tantalizing challenge to the field of botany or just plain gross?

Fresh tomatoes are an exceedingly nutritious food, full of vitamin C and antioxidants. Some studies show that they may lower the risk of certain types of cancer. Tobacco, on the other hand, is full of known carcinogens. Just to read the warning labels on tobacco products is enough to cause severe trauma. Regarding health, the two plants couldn't be more dissimilar.

Nevertheless, in the episode "E-I-E-I-(Annoyed Grunt)" (the expression "Annoyed Grunt" in *Simpsons* episode titles is used to designate Homer's familiar sound of exasperation, "D'oh!"), Homer manages to find common ground (or loam as the case may be) between the two species. It's a curious case of topsoil to ashes, dust to snuff, when the Simpsons move to Grandpa's old farm and try their hand at growing crops. At first Homer doesn't have much of a green thumb—nothing he sows will even sprout—until he decides to apply the substance that boosted the "Amazing Colossal Man" to record heights. His secret ingredient makes his thumb not only green but glowing; it's plutonium shipped to him by Lenny. Soon the farm is blessed with a healthy production of what appear to be tomatoes. Well, maybe *healthy* is not the right word, given that slicing open the red fruit reveals a brown, bitter, ultra-addictive interior loaded with dangerous doses of nicotine.

Realizing that the plant's addictiveness implies a certain commercial potential, Homer dubs the plant "tomacco" and sets up a roadside stand. In short order, bushels of the nuclear product sell like hotcakes—or should we say, like "yellowcake." Everyone who passes the stand wants to try a sample, even little Ralph Wiggum, who reports that it "tastes like Grandma." Once customers have tried a single helping, the nicotine kicks in and they beg for more and more.

Soon the Laramie cigarette company (a fictitious corporation mentioned in several episodes) is interested in marketing Homer's product, particularly because it's legal to sell kids *tomacco*, but not tobacco. They try to negotiate a $150 million contract, but Homer demands a deal-breaking $150 billion instead. Laramie bolts, and later unsuccessfully tries to steal one of the plants. Eventually the entire tomacco crop is devoured by nicotine-crazed farm animals, leaving Homer with nothing left to show for his agricultural efforts.

Although tomacco has since disappeared from the series, it has amazingly popped up in the real world, a case of life imitating art. Inspired by the episode, Rob Baur, an operations analyst for a water treatment plant in Oregon, has grown tomato plants with some of the features of tobacco, including a trace of nicotine content. The method he used, grafting, is a time-tested way of producing hybrids that has nothing to do with Homer's approach.

As any biologist would recognize, fertilizing plants with plutonium would not make them take on other crops' characteristics. Plutonium is a hazardous radioactive substance that is toxic even in minute quantities. It does not exist naturally and is produced and stored under extremely guarded conditions.

Exposure to nuclear radiation can destroy cells or cause cancer. It can also create mutations—changes in the genetic material of a cell or group of cells. Only if such alterations occur to reproductive cells could they potentially be passed on to progeny and possibly manifest themselves as variations in function or appearance.

The bulk of mutations are caused by genetic copying errors during the process of cell division. Some mutations stem from radiation (usually from naturally occurring radioactive sources), chemicals (called mutagens), viruses, and other sources. Nature has erected many firewalls to block the results of unfavorable mutations. For one thing, the body has highly effective repair mechanisms for many forms of genetic damage. Also, because chromosomes are paired, the body has two copies of each gene. As we've discussed, even the Y chromosome itself has several backup copies of its critical genes. Assuming that a mutation is recessive, an organism will favor the healthier version of the gene during reproduction. If instead the mutation is dominant and harmful, it is usually quickly eliminated from the gene pool. Some mutations make no difference at all. Harmless neutral mutations that have no effect on organisms can build up over generations. In extremely rare occasions, a mutation turns out to be beneficial, causing offspring to have a characteristic that improves their survival and ability to reproduce themselves. For example, a mutation could offer greater resistance to a deadly disease. Through the process of Darwinian natural selection, these helpful variations are favored over time and can slowly lead to the evolution of new species.

Exposing tomato seeds to plutonium would be an extremely unlikely way of creating a crop of hybrids. The chances that the genetic material in many different seeds would mutate in just the right way to produce tobaccolike characteristics such as nicotine would be astronomically low. And radiation can't rip genes from one plant and insert them into the cells of another plant. Rather, such genetic modification would need to be performed in a far more controlled situation.

Genetic modification of crops has become, in recent years, a rather controversial issue as it has shifted from the farm into the laboratory. Farmers have used cross-pollination techniques for more than a century to build hardier plants with greater resistance to blight or to have other favorable properties—for example, transferring genes from rye into wheat chromosomes. With the introduction of methods from molecular genetics, modification has become much more precise and has thereby stirred up fears of creating harmful new variations. Foods with genetically modified ingredients have come to be known informally as Frankenfoods.

Grafting, the technique Baur used to make tomacco, is another traditional method in horticulture for blending plant properties that long predates molecular genetics. It involves splicing together the lower part of one plant, including its roots, and the stem, flowers, leaves, and/or fruit of another. After cuts are made, the two parts are carefully positioned together in a manner that permits the free flow of water and nutrients. They are then secured in place until growth occurs and they successfully merge into a single plant. The result is a combination known as a graft chimera or graft hybrid.

For successful grafting, the two original species need to be a reasonable match. Baur realized that tomatoes and tobacco, belonging to the same plant family, had enough compatibility to fit the bill. He recalled a 1959 study in which researchers reported successful crossbreeding of the two species and wondered if the *Simpsons* writers had read the same account. So he went ahead and grafted a tomato plant onto tobacco roots.

Baur's experiment bore fruit—just one, at first. When the fruit was tested, it didn't have any detectable nicotine. The leaves were also tested, however, and they did indeed have some nicotine.

Hence the tomacco plant met the criteria for a true graft hybrid; it had some features from both species. Baur has not marketed his product, so don't expect to find ketchup-flavored nicotine patches in your local pharmacy.

As years of experience have shown, genetic engineering, grafting, and other horticultural techniques appear immensely more effective than radiation in producing hybrid plants. What about the animal kingdom? Could radiation create zoological anomalies, such as three-eyed fish? Let's take a dip into Springfield's "pristine waters" and see what we might dredge up.

3

Blinky, the Three-Eyed Fish

Since the dawn of civilization, water has served multiple uses, from quenching our thirst to washing away our grime. The industrial revolution introduced applications such as supplying steam engines and preventing machinery from overheating. It also created new kinds of pollution that devastated many streams and rivers for centuries, inspiring songwriter Tom Lehrer's 1960s lyrical description of brushing your teeth and rinsing with "industrial waste."[1]

In June 1969, the Cuyahoga River running through Cleveland, Ohio, actually caught fire, likely due to the ignition of an oil slick on its surface. The fire burned for thirty minutes before it was doused. The incident stirred up public outrage against water pollution. A *Time* magazine article described the ghastly conditions of the Cuyahoga: "No Visible Life. Some River! Chocolate-brown, oily, bubbling with subsurface gases, it oozes rather than flows."[2]

Public outrage about the Cuyahoga fire and other examples of industrial pollution became a rallying cry for environmental reform, inspiring the creation of the U.S. Environmental Protection Agency in 1970, passage of the Clean Water Act in 1972, and other related measures throughout the years. In many places, new standards have resulted in dramatically improved water quality. While in no urban area is it advised to scoop up a cup of river water and drink it directly, at least many fish (or, as Homer calls them, "unprocessed fish sticks") have returned to swim and frolic.

Shockingly, given the success of the Clean Water Act, there are still some shortsighted industrialists who try to circumvent the regulations. Holed up in their luxurious mansions, well stocked with spring water from remote mountain sources and protected from thirsty intruders by ferocious baying hounds, they cackle evilly when reading in the newspaper about disgruntled environmentalists. To them, the sound of clinking gold doubloons is much more melodious than the laughter of children playing in a sparkling clear stream.

Could Homer's boss, C. Montgomery Burns, be such a character? Ask his awestruck personal assistant, Wayland Smithers, for a character reference and you'd hear nothing of the sort. Google him under "googly-eyes." However, episodes such as "Two Cars in Every Garage and Three Eyes on Every Fish" paint a more sinister story.

As that episode begins, Lisa and Bart are fishing downstream from the Springfield nuclear power plant and manage to hook a bizarre-looking fish with three eyes. Observing them, Dave Sutton, an investigative reporter trawling for a story, discovers, like the British, that fish and newsprint make a winning combination. Sutton publishes a piece critical of pollution from the plant, which draws the nuclear inspectors to that facility for the first time in decades. There they discover abominations such as gum being used to seal a crack in the cooling tower and a plutonium fuel rod acting as a paperweight. Burns tries to bribe the inspectors, but they are too vigilant and honest. Instead, he decides that the best way to change matters is to run for governor.

Successful political candidates must find ways of addressing their weaknesses, often by framing them in the best light. For Burns, his principal disadvantage has a fishy smell and three glaring eyes. He desperately needs to scoop it from the garbage bin of tabloids and hang it like a trophy. He launches a clever ad campaign, starring an actor playing Charles Darwin, a three-eyed fish named Blinky in a fishbowl, and him. On camera, Burns asks Darwin to explain his theory of natural selection. Based on this explanation, Burns asserts that Blinky has an evolutionary advantage over other fish; in fact it is a "superfish." The campaign effectively combats the fish story and places Burns in the lead.

What Burns and Darwin fail to point out to their gullible audience is that natural selection requires that successful varieties maintain an advantage over others in survival and reproduction. This typically takes many generations to establish. If Burns were scrupulous, he'd examine three-eyed fish over time and see if their ocular or other characteristics allow them to elude predators, more quickly identify food sources, protect their eggs (which only some types of fish do), and so forth. If they don't, then surely the variation would dwindle in population over time, bested by more conventional fish.

Burns purports in the commercial that three-eyed fish are tastier. If that were the case, humans might bolster the stock of such a variety by raising them in protective fish farms, giving them a kind of artificial advantage over less appetizing types. However, Burns's assertion is put to the taste test when he is invited to the Simpson residence for dinner; Marge serves him three-eyed fish, and he spits it out. The media snap photos of Burns's obvious disgust, thus sinking his campaign.

In the world beyond Springfield, three-eyed fish are rarely in the news. Perhaps some of you recall the three-eyed haddock of 1927, featured in the rotogravure section of the *New York Times* on October 16 of that year. The picture had the caption "The oddest fish in the sea . . . a haddock caught off Boston, which was found to have three perfect eyes, the third in the middle of the head."[3]

Dr. E. W. Gudger of the American Museum of Natural History

saw the photo, read an earlier announcement in the *New York Herald-Tribune*, and found the whole thing a bit fishy (pardon the pun). Casting his line of inquiry, he requested additional photos from the *Times*, which they graciously provided. The extra images likewise showed a haddock with a third eye similar to the other two, albeit set farther back on its head. When Gudger tried to snag the actual fish, however, he had no luck. Apparently it had been snapped up by a collector before a scientist had a chance to examine it.

Gudger recalled exceedingly few examples of three-eyed fish throughout history. Those inspected by scientists all turned out either to be malformed embryos or clever fakes. In the former case, these were "double-headed monsters," essentially conjoined twins in which the third eye was shared between the two heads. He could not find any definitive cases of three-eyed embryos that survived into adulthood.

As for the hoaxes, in 1910 Professor Alexander Meek exposed a three-eyed fish scam at the North Shields market in Northumberland, England. Meek came upon the hoax when a specimen was delivered to him for inspection and, upon dissection, he found that the third eye was completely detached from the others as if it had been planted. According to his report, "I noted a small but not at all prominent traverse cut behind the important looking third eye, but that did not prepare me for finding the eye in question quite loose in a cavity behind the normal right eye. It was not connected with anything inside of the head."[4]

The odd part was that several days after Meek's dissection, a man walked into his office in the Fisheries Commission and inquired if anyone had ever caught a three-eyed fish. After Meek informed him about the fake, the man broke down and confessed that he had planned out the whole elaborate ruse. Through practice, he had learned how to cut fish open, insert an extra eye from another fish, and seal the original back up so carefully that not even an expert fisherman could tell that something was amiss. He had planted a few of these fakes at the North Shields market, apparently to see if anyone would notice.

Curiously, all this happened just two years before the perpetration of a much more famous hoax, the excavation of the fraudulent "Piltdown Man." Like the three-eyed fish from Northumberland, Piltdown Man was a deliberate assemblage of parts designed to trick scientists into thinking that they had found a new type of creature. The hoax took place in an age when paleontologists were engaged in an extensive search for the "missing link": the immediate precursor of *Homo sapiens*, but with some apelike characteristics. The idea was that variations of this being, through mutation or another means, were favored over time and evolved into modern, large-brained, fully upright human beings. (Refer to Homer's courtroom behavior in the episode "The Monkey Suit" to gain an inkling of what that missing link could have been like.)

In 1912 Charles Dawson discovered the first of two Piltdown skulls beneath a quarry located in Sussex, England. Possessing a full manly brow but a rudimentary apelike jaw, it seemed tailor-made to complete the geological record of human evolution. In fact, as experts established in the late 1940s and afterward, the skull was actually less than fifteen hundred years old at the time of excavation and had almost certainly been planted by someone. Historians have considered many possible perpetrators, with Dawson (who died in 1916) being a leading candidate. Fortunately, the Northumberland fish-mutilator confessed, or experts could have been wondering about that one, too.

The chap who stitched together Gudger's three-eyed haddock, on the other hand, never stepped forward. However, in 1928 Gudger came across a report by a weathered fisherman (somewhat reminiscent of *The Simpsons*' nautical character, the Sea Captain) that seemed to confirm his suspicions. The fisherman relayed an account of

> an old feller [who] was a great one for whittling, real handy with a knife. Well he was working on the head of a haddock, real careful like, and when he got through, he brought out a fish eye from his pocket and slipped it in the hole, just as neat as you please. Never saying a word, he drops the three-

eyed haddock back with the other fish, and the next day, folks were coming from far and wide to see the latest wonder of the world, the three-eyed haddock.[5]

Since the time of Meek and Gudger, there hasn't been much written about three-eyed fish, except for cultural references to Blinky and speculation about deformities due to nuclear radiation. As Dr. Anne Marie Todd of San Jose State University pointed out, Blinky serves as a visual reminder of the clash between official polemics and environmental facts on the ground, even if three-eyed fish don't really swim around the rivers near power plants. Todd remarked:

> This episode condemns the manipulation of political and economic power to disguise ecological accountability, and shift blame for environmental problems. The show comments on the lack of adherence to safety standards for the plant, and criticizes the apathetic acceptance of unforced environmental inspections. Finally, this episode explicitly criticizes media spin-doctors who distort the impacts of ecological degradation caused by wealthy corporations like the nuclear power plant.[6]

Indeed when the public grapples with issues concerning nuclear power, the possibility of mutations that cause deformities often leaps to prominence. Images of animals with extra eyes, multiple heads, and so forth are visually arresting. Evidence appears to indicate, however, that this focus misses the mark. Although there are many serious questions about nuclear energy, including the costs involved in building and decommissioning plants, the problem of nuclear waste disposal, and the danger of fissile material falling into the hands of terrorist groups, there has not been any statistical increase in inherited abnormalities in the proximity of functioning nuclear facilities. The 1986 Chernobyl disaster in Ukraine, the worst in the history of the nuclear industry worldwide, is another story, causing horrific damage to the health and environment in its region. It had far greater impact than the Windscale fire in England and the Three Mile Island accident in the United States, the two best-known prior nuclear accidents.

Chernobyl's design was particularly poor, with the cores of each of its four reactors composed of graphite, a flammable material, and with an inadequate backup system in case of fire. Each graphite structure was full of slots that housed the nuclear fuel elements powering the reactors. As with all commercial reactors, these elements produced energy through the process of fission. While normally this process is carefully regulated at a power plant, in Chernobyl the lack of proper safeguards led to a fire and the release of hazardous radioactive materials that spiraled out of control.

Fission is a splitting up of large atomic nuclei: the sets of protons (positively charged particles) and neutrons (neutral particles) that constitute the cores of atoms. When a fissile material such as uranium 235 is bombarded with relatively slow neutrons, each nucleus splits into several fragments, producing energy and more neutrons in the process. These neutrons, in turn, induce more fissile material to divide, causing a chain reaction. The byproduct is various isotopes (variations of elements with differing numbers of neutrons), some of which are radioactive. While the reactor is running, some of the heat produced creates steam that runs a turbine to generate electricity. The electricity produced can supply communities with a steady source of power.

Ordinarily, control rods, placed in between the fuel rods, modulate the process by absorbing neutrons. Inserting and removing these control rods as needed helps make sure the reactor runs efficiently and not out of bounds. Also, coolant (cold water) bathes the rods, preventing them from becoming too hot. In the case of the Chernobyl disaster, however, too many control rods had been removed from the core of one of the reactors at a time when there was too little coolant. The water that remained turned into steam, and the core temperature began to rise out of control. Fuel rods shattered, and the hot uranium mixed with the steam. Enormous pressure built inside and the reactor's top lifted off, allowing air to enter the chamber. Unlike reactors in other parts of the world, it didn't have a containment vessel. The hot graphite, mixed with oxygen, generated carbon monoxide gas and caught on fire. Billows of radioactive smoke, containing fissile and waste products, spread throughout the community, contaminating farms and towns for hundreds of miles. Although the reactor was shut down, filled with liquid nitrogen to cool it off, covered with sand to put out the fire, and later encased in a thick concrete tomb, horrendous damage had already been done.

When radioactive material is spread throughout a wide region it can boost the rate of cancer, cell death, and mutation well beyond what would be expected due to natural radiation levels and other causes. Experts estimate that Chernobyl caused thousands of deaths due to cancer, radiation poisoning, and other effects. The *Guardian* reported in 2001 that the mutation rate for the children of the workers who helped clean up the disaster was 600 percent higher than normal. This jump in mutations was determined through genetic testing, not through a profile of symptoms, because the DNA changes were not large enough to produce deformities, at least for that generation. Hence even a tragedy the magnitude of Chernobyl, while leading to innumerable deaths, did not create abnormalities such as malformations in people or three-eyed fish.

Fear of any repeat of such a calamity is one reason nuclear safety has continued to be a major public concern. The world's environment cannot afford another Chernobyl. Therefore, while the likes of Blinky may not be seen in nature, even in the waters near reactors, his grotesque image well captures our deepest fears about nuclear perils.

Present-day concerns about the perils of radiation are a far cry from its image a century ago, when it was seen as a cure-all. When Burns touts radiation's benefits, his message is a throwback to the bad old days when radium, a naturally radioactive element, was largely mishandled due to ignorance about its dangers. It was even included in "health tonics" that were supposed to give users increased vitality and a "healthy glow." Indeed Burns himself has such a glow, but whether it's healthy or not is a different matter.

4

Burns's Radiant Glow

Every city has its great establishments of victuals and libations where prominent thinkers converge. New York has the legendary Algonquin, renowned for its Round Table. Vienna has its Café Central, Café Sacher, and innumerable other coffee shops, where philosophers exchange retorts over tortes. Paris spans the gamut from tony Maxim's to tiny bistros. Although not in the same league as New York, Vienna, and Paris (and, according to its snobbish residents, not even neighboring Shelbyville), Springfield can boast of its spartan but comfortable Moe's Tavern, serving cold, refreshing Duff Beer.

Just as in the days of George Bernard Shaw and Oscar Wilde, the witty banter at Moe's (bon mots, perhaps) is often preceded by a few drinks. The more inspired the individual, the more unusual the drink. In the episode "The Springfield Files," Homer decides to be very creative and try a new brew called "Red Tick Beer." Soon his level of inebriation, if not creativity, has reached Ernest Hemingway proportions. A Breathalyzer test confirms this, and Moe insists that Homer walk home.

At least one of the routes from Moe's Tavern back to 742 Evergreen Terrace (where the Simpsons live) winds through a deep, dark forest. Homer, in his drunken state, decides to take that path. While in the midst of the woods, he sees in the distance what appears to be an odd-shaped, wide-eyed glowing green alien. Naturally Homer is terrified. When the being tries to reassure Homer in a strangely soothing mellow voice, Homer screams and runs away. Finally arriving home, Homer is dismayed that nobody believes that he encountered an alien. His intoxication at the time of his sighting makes his story all the less credible. Who in Springfield would trust him enough to help him conduct a full investigation? Could his needed assistance come from someplace beyond?

This particular episode is what in television vernacular is called a "crossover." The term has a variety of meanings. In biology, crossover occurs during the process of cell division when two paired chromosomes, one from each parent, each break at a particular point and exchange segments of genetic material. This stirs up the pot of alleles (different forms of the same gene), offering the prospect of new combinations of traits. Along with mutation, it is one of the principal sources of biological variation—changes that can be neutral, negative, or positive. In the best-case scenario, the mixing produces novel characteristics that heighten the child's environmental fitness. Favorable variations, over the long run, further the process of evolution. In television, on the other hand, crossover occurs when characters from one series appear on another, usually within the same network. In the best-case scenario, the mixing produces novel plot devices that increase the show's ratings.

The crossover in this case involves Mulder and Scully, the ace paranormal investigators and FBI agents from the popular 1990s series *The X-Files*, played by David Duchovny and Gillian

Anderson. Learning of Homer's encounter, they visit his house and attempt to unravel the mystery. Homer leads them to the woods where he saw the glowing being. Suddenly, someone emerges from the bushes. It turns out to be just Grandpa, who has been lost there for days. Exasperated, Scully departs and Mulder eventually follows after delivering a long speech about the mysteries of the universe.

Only during a return trip to the woods, this time without any FBI agents to help them, do Homer and the townspeople of Springfield learn the truth. They are joined by Leonard Nimoy (voice-performed by himself), who is similarly eager to resolve the seemingly inexplicable. When the creature appears once more, conveying a message of love, Lisa illuminates it with a flashlight. They discover that the "alien" is actually Mr. Burns, who has been receiving medical treatments that have altered his appearance and affected his behavior. Eye drops have enlarged his pupils, chiropractic treatments have changed his posture, a vocal chord procedure has modified his voice, and painkillers have artificially bolstered his mood. The reason this treatment sounds so bizarre is that his physician is the total quack Dr. Nick Riviera.

What about the eerie luminescence? Burns explains that his lifelong nuclear profession has given him a "healthy green glow." For some reason, not explained in the show, the glow appears only during his nocturnal jaunts into the woods after his medical treatments.

In the history of bogus medical treatments, there have been many relationships like that of Mr. Burns and Dr. Nick. Wealth attracts quackery like wolves to fresh meat. One of the most notorious examples of a charlatan preying on an entrepreneur regards the not-so-healthy effects of radiation as delivered through a radium-infused "health tonic." The only positive thing to emerge from the incident was greater public knowledge of radiation's dangerous effects.

Radium, the eighty-eighth element in the periodic table, was chemically identified in the late 1890s by Marie and Pierre Curie, working in a Paris laboratory. It was one in a series of important discoveries at the time regarding the properties of radioactivity. In

1895, the German physicist Wilhelm Roentgen discovered invisible emissions, called X-rays, that came to play a critical role in imaging the skeletal structure of the human body. The following year Antoine Henri Becquerel found that uranium salts gave off unseen rays that would expose covered photographic plates. Thus uranium became the first known radioactive substance. Today we know that radioactive materials mainly emit three different types of radiation: alpha particles, also known as the helium atom nuclei; beta particles, or electrons; and gamma rays, invisible high-energy light. To understand the cause of the emissions from uranium salts, the Curies set out on an intensive search, culminating in the isolation of two elements in 1898: polonium (named after Marie's native Poland) and radium. Each of these elements, they found, were extremely radioactive.

Excited by the properties of invisible radiation, a kind of "radiomania" swept the world. Marie Curie made it part of her personal mission to find use for radium in the diagnosis and treatment of medical ailments. Patients with tumors were gratified to see that an application of radium could help shrink them. Today, radiation therapy, under far more controlled conditions, is still used to help reduce malignancies (cancerous tumors), especially when surgical removal and other forms of treatment are impossible.

Back in the early part of the twentieth century, however, no one had any inkling as to the extreme perils of radium poisoning. With hope for rejuvenation, naive members of the public bathed in radium spas and drank radium tonic. Meanwhile, radioactive ores were mined with few precautions. All this came to a close, however, with the publicity surrounding the horrible death of a prominent figure from a quack "remedy."

Eben Byers was a physical powerhouse and industrial giant. An exceptional athlete with a strong build, at the age of twenty-six he won the 1906 U.S. Amateur Golf Championship. During the 1910s and 1920s, he was a well-known socialite and wealthy tycoon, heading the A. M. Byers Iron Foundry while owning estates in New York, Pittsburgh, South Carolina, and Rhode Island. When newspapers in March 1932 announced his untimely death from radium poisoning and revealed the dreadful manner of his demise, not only were friends and family members shocked, the general public was mortified as well. As Roger Macklis, an expert on radiation oncology and radiomedical quackery described:

> When Byers died, his shriveled body must have been barely recognizable to friends who had known him as a robust athlete and ladies' man. He weighed just 92 pounds. His face, once youthful and raffishly handsome, set off by dark, pomaded hair and deep-set eyes, had been disfigured by a series of last-ditch operations that had removed most of his jaw and part of his skull in a vain attempt to stop the destruction of bone. His marrow and kidneys had failed, giving his skin a ghastly cast. Although a brain abscess had rendered him nearly mute, he remained lucid almost to the end.[1]

Who was the unscrupulous "Nick Riviera" who duped this unfortunate gentleman with a deadly bogus cure? The blame falls squarely on quack doctor William Bailey and his poisonous potion called Radithor. Bailey had a history of producing fraudulent medication, including a treatment for male impotence that included strychnine as its active agent. In 1918, he was fined more than $200 for falsely advertising this dangerous drug as a cure-all, but that didn't seem to deter him. Then in 1921, after Marie Curie completed a tour of the United States promoting the possibilities of radium, Bailey became extremely interested in her work. He began to investigate and develop different kinds of radium treatments, including radioactive pendants that could be worn on various parts of the body to promote healthy metabolism. Finally, he concocted his best-selling product: a solution of radium extract in distilled water. Marketed as a revitalizing tonic to physicians around the country (along with a special promotional discount), Radithor was a smash success that proved extremely lucrative for its inventor.

Even when mounting evidence suggested that even tiny doses of ingested radium could prove lethal, Bailey insisted that his potion was harmless.

Byers began his use of Radithor in 1927, when a physician recommended it as a treatment for injuries he had suffered during a fall. At first, the radium elixir made him feel fresh and vigorous. Enthusiastic about the medicine, Byers began to drink more and more. The radium seeped into his bones and began to whittle them away. By the time he stopped using the drug it was too late; the damage was irreversible.

Shortly before Byers died, a savvy radiologist examined him and diagnosed his condition as radium poisoning. The Federal Trade Commission, already investigating Bailey for fraudulent practices, shut down his business. Surprisingly, Bailey managed to weasel his way into further scams and felt only minimal repercussions for the great harm he had inflicted. Nevertheless, the scandal led to a toughening of drug regulations, particularly restrictions on the sale of radioactive medicines.

Today, physicians take great care in minimizing patients' exposure to potentially harmful forms of radiation, unless it is a component of a specific type of treatment critical to their health. Even low-level radiation has come under increasing scrutiny. Although in popular media radiation exposure has come to be associated with either gruesome deformities (three-eyed fish) or awesome superpowers (as in Bart's favorite comic, Radioactive Man, who acquired amazing strength through a nuclear blast), the reality is far more likely to involve unseen internal changes that in unlucky cases prove deadly over time.

Likewise, among all the risks of radiation exposure, we cannot count glowing skin. Despite Mr. Burns's report, his lifetime exposure to nuclear emissions could not have made him light up like a glow-in-the-dark watch, unless he actually covered himself head-to-toe with the radioactive ink used by dial painters in the early twentieth century. But who knows what other maladies radiation has inflicted on Springfield's leading entrepreneur, given his plant's appalling safety record?

Unquestionably, Mr. Burns is not a well man. Even with all his wealth, he cannot postpone bodily decay forever. Differentiated (specialized) human cells cannot divide forever; biologists have discovered that faithful replication can take place only a finite number of times before imperfect copies are made. Intriguingly, stem cells (unspecialized cells before differentiation) and tumor cells can divide indefinitely, at least in laboratory situations. Stem cell researchers are currently trying to understand the distinction, and eventually perhaps will even learn how to reverse aging. Until such a breakthrough is made, the limitations of how many times the cells in our bodies can produce healthy copies of themselves mandate that life is not infinitely renewable.

Someday, perhaps, as Eric Drexler suggested in his seminal book *Engines of Creation*, scientists will design tiny robotic agents of molecular size (on the order of one nanometer, or 39 billionths of an inch, known as "nanoscales") able to wander the body and repair cell damage. This could lead to a dramatic extension of the human lifespan.

These nano-agents would need to be dexterous and savvy, able to navigate tight channels and render instant judgments. In a way, they would be like miniature versions of faithful nonunionized employees carrying out arduous tasks under demanding circumstances. If they could be injected into the frail body of Mr. Burns, perhaps they could rejuvenate him. Any volunteers? Exx-cellent.

5

We All Live in a Cell-Sized Submarine

The Simpsons' often disturbing level of dysfunctionality suggests that they could well use a shrink. The one time they were shrunk, however, didn't seem to help matters. Sadly, their problems just seemed to loom larger.

In the Treehouse of Horror XV segment, "In the Belly of the Boss," the town's leading scientist, Professor Frink, constructs a "shrink-ray machine" to reduce a gigantic vitamin capsule (packed with a lifetime's supply of nutrients) to conventional size so Mr. Burns can swallow it. Just before the capsule is miniaturized, Maggie crawls into it and is shrunk and swallowed along with it. It's a bitter pill indeed for Marge and Homer when they've lost their baby daughter to the stomach of Homer's boss.

Surrounded by the capsule, Maggie has a bit of time before digestive juices eat her cradle away. Accordingly, Homer, Marge, Bart, and Lisa decide to attempt a rescue. In the style of the 1966 film *Fantastic Voyage*, they board a submarine and are shrunk down to something like one micrometer (less than 1/25,000 inch) in size. Then Frink injects them into Burns's body, and off they go in search of Maggie.

Frink radios the crew detailed instructions, which Homer in typical fashion completely ignores. After Homer foolishly presses all the buttons on the ship's console, the ship gets stuck and the Simpsons need to wander outside to dislodge it. While they are in the bloodstream, white blood cells attack Marge's clothes but amazingly know exactly when to stop before revealing too much. The family's repair is successful, and they manage to find and rescue Maggie. Unfortunately, with Maggie's additional weight the craft is overladen, and one crew member must be left behind. Homer is elected, and the rest manage to escape. Once the shrink-ray's spell wears off, the Simpsons revert to normal size. Homer assumes his full girth while still inside Burns, turning the pair into a kind of two-headed hydra. The segment closes with the two of them still conjoined, singing and dancing to the Frank Sinatra chestnut "I've Got You under My Skin." It's doubly enough to make the skin crawl.

From classic fantasy epics such as *Gulliver's Travels* and *Alice in Wonderland* to modern film comedies such as *Honey, I Shrunk the Kids*, miniaturization has long been a favorite topic of speculation. The human race's natural variability in height has inspired fictional accounts of colossal giants and tiny sprites. Could people grow taller than a steeple or become smaller than a thimble? Might there be unknown lands where Goliaths scoop up livestock with their cupped hands and chomp on cattle a dozen at a time? Or conversely, places where Lilliputians scurry like ants at the sight of an ordinary spider?

Although it's a marvelous concept, miniaturization of people would be an exceedingly tough order for science. To shrink someone to the size of a dust mite would require either reducing his quantity of cells, making his cells considerably smaller, and/or compressing down the space within the molecules and atoms that form the cells. Each of these steps would almost certainly be impossible to achieve without destroying the individual.

For instance, eliminating enough cells from someone's body to make him much smaller would render most of his organs inoperative. Hearts need a certain amount of muscle tissue to function properly, and brains require a minimum number of neurons. In diseases where the brain shrinks in size, such as late-stage Alzheimer's, for example, a person loses a great deal of his or her cognitive functioning. And even that amount of shrinkage is far less than what would be required to miniaturize a person. So we can rule that option out.

A second choice would be to preserve an individual's number of cells, but shrink down the cells themselves. Such a scheme also couldn't possibly work. To reduce someone in height from five feet down to several ten-thousandths of an inch would require each human cell, ordinarily possessing hundreds of billions of atoms, to become so small that it couldn't contain a single atom. That's because a reduction in diameter by a certain factor implies a decrease in volume by that factor cubed. If Earth were shrunk that much it would be less than the size of a house and clearly couldn't contain the same amount of material. The same thing applies to cells reduced to subatomic scale; they certainly couldn't harbor the macromolecules (large molecular chains)—proteins, genetic material, carbohydrates (sugars and starches for fuel), and so forth—that they would need to function.

In 1998, the National Academy of Sciences held a conference examining the critical question "How small can a free-living organism be?"[1] Participants at the meeting considered the minimum ingredients for viable primitive cells. What would be the bare requisite for them to replicate, maintain their shape and content, and engage in the basic biochemical processes that we associate with life? The consensus was that the smallest viable cell would require approximately 250 to 450 genes and 100 to 300 species of protein.

If the cell possessed 1,000 copies of each kind of protein, then its minimal diameter would lie somewhere between 200 and 300 nanometers, or on the order of 1/100,000 of an inch. Note that these parameters concern the simplest possible kinds of cells; the cells of advanced organisms such as humans are far more complicated.

Biologists classify cells into three basic types, based on their degree of complexity. The simplest kind of cell, called prokaryotes, includes single-celled organisms such as bacteria and cyanobacteria (also called blue-green algae). The most common type by far, prokaryotic cells are organized in a very basic fashion. First, they have a single DNA molecule constituting a long, coiled chain of genes wrapped up in a dense central region of the cell called the nucleoid. Nothing physically separates the nucleoid from the rest of the cell, allowing for close contact between the DNA strand and scattered centers, called ribosomes, where RNA assembles amino acids into proteins. Other essential contents include the proteins themselves, carbohydrates, and fats. Surrounding the cell's interior is a thin layer called the cell membrane that protects the cell by selectively allowing only certain types of material to pass through. Typically, but not always, the prokaryote is housed within a rigid cell wall, a kind of fortress that helps maintain its shape and further guards it against unwanted intruders.

Another basic type of cell, a newer category discovered only in the 1970s, is known as archaea. Like prokaryotes, these are simply organized and do not have much in the way of internal structure. However, their composition and outer layers are different, permitting these organisms to thrive under extremely harsh conditions, such as hot springs, corrosive chemicals, and thermal vents (cracks on the ocean floor where magma bubbles up from Earth's fiery interior). If Burns wanted to breed pets in his reactor's steam pipes and didn't want to pay for upkeep, archaea might make a promising choice.

The third cell category, eukaryotes, are considerably larger and more complex than the other two types. Typically more than a thousand times greater in volume than prokaryotes, eukaryotes are organized into specialized substructures, called organelles, which perform a variety of different tasks. If prokaryotes are like tiny Kwik-E-Marts where a mixture of items is immediately accessible, eukaryotes are akin to big-box stores, where items are grouped into specific departments, each with its own specialty. For example, the cell nucleus, akin to the management office, houses the cell's principal genetic material. (Note that a cell nucleus should not be confused with an atomic nucleus; they are vastly different things both in size and function.) Mitochondria, functioning like furnaces, produce the cell's energy through biochemical processes. They also house their own groupings of genetic material, making them something like cells within cells. Lyosomes break down waste products, the endoplasmic reticulum folds proteins into functional shapes, and the Golgi body finishes the protein-folding process, assembles sugars into starches, and links proteins with sugars to form what is called glycoproteins. Additionally, they include a protein-based structure called a cytoskeleton, a jellylike substance called cytoplasm, numerous ribosomes, a cell membrane, and a number of other components. Advanced beings, including humans, are made of various kinds of eukaryotic cells, allowing a much more sophisticated level of functioning than single-celled organisms such as bacteria exhibit.

Because of their incredibly detailed structures, the smallest eukaryotic cells, such as the red blood cells of mammals, likely couldn't become more compact than their typical size of 8 micrometers (about 1/3,000 inch) in diameter. Hence, the size of functioning adult human beings has a lower boundary due not just to the minimal number of cells, but also to the minimal requirements of eukaryotic cell structure.

Finally, let's consider a third potential means of miniaturization: reducing the size of atoms themselves. If shrink rays were to make everything smaller, including inanimate objects, they would need to shrink down the atomic building blocks that constitute all types of materials on Earth. Stable atoms, however, have a limited spectrum of sizes governed by the principles of quantum mechanics and could not be compressed without significantly altering their properties.

Quantum mechanics was developed in the early decades of the twentieth century as a means of explaining certain mysteries concerning the relationship between matter and radiation. In the nineteenth century, the brilliant physicist James Clerk Maxwell showed that light is an amalgamation of electric and magnetic fields, known as electromagnetic radiation. A field is a measure of the strength and direction of forces in various parts of space. When an electric charge oscillates, it generates electric and magnetic fields that are at right angles to each other and move through space as a wave. We observe that phenomenon as light. The rate of these oscillations determines what is called the frequency of the light.

The discovery of X-rays proved that light takes invisible as well as visible forms. In the visible case, light's frequency manifests itself as color. The highest frequency of visible light is violet, and the lower frequency is red, with the other colors forming a rainbow in between, called the visible spectrum. The full electromagnetic spectrum, however, includes a wide range of invisible forms of radiation, from low-frequency radio waves, microwaves, and infrared radiation to high-frequency ultraviolet radiation, X-rays, and gamma rays—the highest frequency of all.

It's easy to observe the visible spectrum by holding up a prism or a diffraction grating (a screen with thin slits) to a light source and observing the result on a screen. Each of these optical devices breaks up light into its spectral components by deflecting each frequency by a different angle. For a common source of illumination, such as a lightbulb, the image on the screen would represent a complete palette of colors. However, for a lamp containing a pure chemical element in gaseous form, such as hydrogen, helium, or neon, only particular colors corresponding to certain frequencies would be seen, as thin lines separated by gaps.

The existence of such fixed spectral patterns represented a great mystery to the physics community at the turn of the twentieth century, particularly because the values of these frequencies followed predictable mathematical formulas. For example, a formula developed by the Swiss mathematician Johann Balmer and generalized by the Swedish physicist Johannes Rydberg predicted exactly where certain hydrogen spectral lines would fall. Physicists were perplexed as to why hydrogen's rainbow has predictable gaps.

In 1900, Max Planck made a major breakthrough with his proposition that energy is quantized—or found only in tiny packets. In examining a phenomenon called blackbody radiation (energy emitted by a perfect absorber of light), he observed that the frequency distribution for a given temperature could be well modeled by a formula that assumes energy values that are multiples of the light's frequency and a physical constant now known as Planck's constant. (A physical constant is a fundamental natural quantity believed always to maintain the same value. Other examples include the speed of light in a vacuum and the smallest quantity of electric charge for a free particle.)

Planck's calculation was akin to estimating the number of coins in a piggy bank by knowing the total amount as well as what denominations coins can be. Let's suppose, for instance, a group is told that the bank has five dollars' worth of coins and is asked to guess how many coins are in it. Someone who thinks that the coins are all pennies would give an estimate different from someone who believes the coins are roughly an even mix of pennies, nickels, dimes, and quarters. A third person who erroneously believes that American coins could come in any denomination, including two-cent and three-cent pieces, would likely make yet another guess. Similarly, positing that photons (light particles) can have only particular energy values gives an estimate for the blackbody frequency distribution different from the assumption that they could have any amount. Planck demonstrated that the former hypothesis yielded the correct distribution—a milestone for modern physics.

In 1905 Planck's hypothesis was further bolstered when Albert Einstein proposed the photoelectric effect. Einstein's Nobel Prize–winning discovery predicted what would happen if beams of light of various frequencies were aimed at a piece of metal and released electrons from its surface. The traditional theory of waves suggested that the more intense (brighter) the beam, the greater the energy it would impart to the electrons and, once freed from the surface, the faster they would move. As Einstein predicted, however, that's not what happens. Instead, the energy transferred from the light to the electrons is conveyed in multiples of fixed amounts: namely, Planck's constant times the frequency. This proved definitively that light is quantized; it comes in discrete photon "packets" rather than in continuous waves.

With Planck's quantum concept well established, in 1913 the Danish physicist Niels Bohr applied it to the mystery of the atom and found a clever way of reproducing the patterns of spectral lines. Bohr's pioneering contributions earned him not only a Nobel Prize (in 1922, the year after Einstein received his), but also a brief mention on *The Simpsons*. In the episode "I Am Furious Yellow," one of Homer's favorite TV shows is preempted by the program *The Boring World of Niels Bohr*. Homer is so upset that he clutches an ice-cream sandwich, aims it at the screen like it's a remote control, squeezes out its contents, and splatters Bohr's image. In contrast to Homer's reaction, most physicists heap nothing but accolades upon Bohr, whose revolutionary ideas shaped the modern concept of the atom.

Bohr's atomic model has several key assumptions. First, in line with experiments by Ernest Rutherford, atoms consist of a positively charged nucleus orbited by negatively charged electrons. We have seen how large atomic nuclei break down in the process of nuclear fission; the type Bohr considered was much simpler. The most basic nucleus—that of hydrogen—is just a single proton.

Bohr's second assumption is that the force causing electrons to orbit is just the electrical force, obeying Coulomb's law that its strength varies inversely with the square of the distances between charges. In other words, as electrons approach a nucleus, the nucleus's force of attraction gets stronger and stronger. Therefore electrons like getting closer if they possibly can.

What, then, holds electrons back from plunging into atomic centers, rendering all matter unstable? Bohr further supposed that an electron's angular momentum (essentially its mass times its velocity times its radius) can take on only discrete values—whole-number multiples of Planck's constant divided by the mathematical quantity two times pi. The specific whole-number multiple—1, 2, 3, 4, and so forth—is called the principal quantum number. By assuming that angular momentum, like energy, is quantized, electrons are forced into fixed orbits with only particular radii.

One more assumption Bohr made allowed him to reproduce the formulas of Balmer, Rydberg, and others for predicting the frequencies of the hydrogen spectrum. He supposed that electrons could jump from one orbit to another, emitting or absorbing a photon in the process. If an electron leaps to a lower orbit, it gives off a photon that carries away the energy difference between the levels. Because energy is proportional to frequency, the light's color depends on how much energy the electron loses. A large leap might give off violet light, for example, and a small one red. Conversely, if an electron absorbs a photon of just the right energy it can move up to a higher orbit.

Just how low can it get? This question is unanswerable if applied to Springfield's mayoral elections; nevertheless it has precise application to atomic electrons. The tightest electron orbit, called its ground state and possessing a principal quantum number of 1, is the absolute minimum. Electrons simply can't get any closer. For hydrogen, the ground state has a radius of approximately 5.3×10^{-11} meters (two billionths of an inch), known as the Bohr radius.

Bohr's rudimentary theory was later supplanted by a fuller approach to quantum mechanics, developed in the 1920s by the physicists Louis de Broglie, Werner Heisenberg, Max Born, Erwin Schrödinger, and others. In its comprehensive form, quantum mechanics posits that electrons can be represented by wave functions, entities without exact locations but rather smeared distributions centered on certain average positions. Even in the revised theory, however, Bohr's basic prediction of quantized electron states has held true. Thus, search as hard as you'd like, you would never find a hydrogen atom with an electron ground state centered in a region smaller than its Bohr radius.

The presence of a minimal atomic radius casts a monkey wrench into our final hope for miniaturization—shrinking down atoms. Atoms simply can't be reduced in size arbitrarily. Throughout the universe, natural elements appear to have standard spectra, indicating there's no such thing as shrunken hydrogen, oxygen, carbon, and so forth—some of the essential building blocks of life's molecules. Thus, in this cosmic democracy, Mayor Quimby's atomic constituents would have just the same breadth, depth, and scope as anyone else's. Universal equality, at least on the microscopic scale, is a firm law of nature.

Thus, the next time the Simpsons are requested to see a shrink or to be positioned in front of a shrink ray, they could well point out that the atoms in their blood are just the same as those of any other law-abiding carbon-based life form, and thereby, according to physical principles, couldn't be crushed into teeny-tiny "atomlettes." Fundamentally, the message "Don't Tread on Me" seems to be written on our cells like a flag. Of various fantastic scenarios, miniaturization doesn't seem to lie within the realm of scientific possibility, at least according to our current understanding of quantum physics.

6

Lisa's Recipe for Life

From the Golem to Frinkenstein (monsters featured in Treehouse of Horror XVII and XIV), one of the specialties of *The Simpsons* is breathing life into the inanimate. Perhaps this is an echo of what the show's writers and artists are doing themselves when they set into motion the characters' interactions on the screen. Creating the *illusion* of life is an age-old form of artistic expression, from Punch and Judy to virtual reality. But what about *truly* creating life and fashioning genuine living beings from lifeless materials? Will humanity ever unravel the secrets of genesis?

Of all the members of the Simpson family, the one with the greatest concern for matters of life and death is Lisa. As a vegetarian and a Buddhist, her solemn vow is to treat all living beings as sacred. The last thing she would want to do is play God and decide which among the creatures of nature should survive and which should perish.

When Lisa, in the Halloween segment "The Genesis Tub," takes on the role of creator and sustainer of an entire miniature civilization, she is placed in an awkward position. Although she is generally supportive of bringing new scientific knowledge into the world and welcomes discoveries that will broaden our understanding, she comes to realize that being a creator is a colossal burden as well as a source of accomplishment.

The segment starts off with Lisa assembling a science project. One of her baby teeth has just fallen out, so she places it in a tub and pours Buzz Cola on it to examine the soft drink's corrosive effects. Bart, as usual, is not particularly supportive. He spitefully gives Lisa a static electric shock, which she passes on to the tooth. Miraculously, the electric jolt sparks the cola-soaked tooth to start growing miniature life forms around it. Through an ultra-rapid evolutionary process, a thriving town full of tiny people emerges. Lisa has created her own world.

By listening to Lisa's voice, the microbe-sized citizens learn English and pick up Lisa's disdain for Bart's antics. They develop a religion that associates Lisa and Bart with divine and diabolical roles, respectively. A miniature Professor Frink invents a "debigulator"—similar in function to a shrink ray—that he uses to reduce Lisa to their size. Worshipped by all the little beings, Lisa is placed on a throne and asked to resolve their deepest theological questions. Meanwhile, Bart, still normal size, takes credit for Lisa's experiment and wins a school prize. Lisa, frustrated by her inability to return to normal size and the failure of her attempts to communicate with the outside world, reluctantly comes to accept her role as the civilization's leader. By creating a tiny race, she has been forced to share its fate and guide its future.

From our previous discussions we must take the idea of miniature people with a nanogram of salt. Human brains are extraordinarily complex—packed with 100 million neurons, each one of which is a vastly intricate cell. How then could microscopic beings the size of mold spores possess anything close to human know-how? Moreover, if terrestrial evolution were to repeat itself and produce something like people, wouldn't it have taken a comparable time scale and yielded beings of similar size?

One thing science has learned in recent years is that the scales of things are usually no accident. From galaxies down to atoms, each of nature's players has a range of proportions determined by the fundamental laws and conditions of the universe. Hence, don't expect to create a galaxy, a star, a planet, or even advanced beings in your kitchen sink—only, perhaps, if you try hard enough, a stunning race of fungus.

Let's set aside the issue of creating tiny people and return to the more realistic question of whether science could bring forth any kind of life from inanimate materials. This long-standing riddle bears strongly on the question of how common life is throughout the universe. The easier it was for life to have emerged on Earth, the greater the chances that it has bloomed elsewhere.

One of the earliest and most frequently cited research projects designed to examine this question was the Miller-Urey experiment, developed in 1953 by the graduate student Stanley Miller under the supervision of the Nobel laureate Harold Urey at the University of Chicago. The experiment attempted to re-create primordial conditions of Earth and see if organic materials necessary for life would emerge. Within a labyrinth of interconnected glass tubes and spherical flasks, Miller combined four different substances known to exist in Earth's atmosphere billions of years ago: methane, ammonia, water, and hydrogen. In a series of cycles, he heated the water until it evaporated, applied electric sparks through the mixture (to simulate primitive Earth's electrical storms), and then cooled the water down again until it condensed. After a week of running the experiment, he tested the mixture using the technique of paper chromatography to determine its composition. Remarkably, he identified a number of common organic substances, including many of the amino acids (such as glycine) that serve as building blocks for proteins.

Since the time of the Miller-Urey experiment, the field of biology has undergone an extraordinary revolution in researchers' ability to produce and manipulate the requisites for life. One of the greatest breakthroughs has been the development of methods for cutting and splicing together various DNA strands to form what is called recombinant DNA (rDNA). These custom-made genetic templates can be inserted into host cells—either bacteria or eukaryotic cells—that can be induced to produce recombinant proteins. A wide range of synthetic proteins has been manufactured in this manner, from synthetic human insulin to synthetic human growth hormone. Thanks to emerging biotechnologies, medical science has been able to develop new life-saving treatments and may ultimately be able to correct a number of genetic disorders.

Even as our understanding of genetics becomes increasingly sophisticated, science is still unsure exactly how simple cells emerged from the primordial organic broth billions of years ago. One researcher who has devoted many years to this issue is the Harvard molecular biologist Jack Szostak (not to be confused with Moe Szyslak). As part of Harvard's "Origins of Life in the Universe" initiative, Szostak has investigated a theory that fatty acids and RNA, when supplemented with certain types of clay, could have assembled themselves into primitive cells, complete with membranes. As Szostak once described his research: "Cell membranes self-assemble under the right conditions. If you sprinkle a little bit of clay into these reactions, it speeds them up."[1]

As a leading expert in biotechnology, Szostak was one of the pioneers of recombinant DNA research and has also studied the role of telomerase, a critical enzyme that protects DNA strands from becoming shorter each time cells divide. Without this enzyme, DNA deteriorates over time. Szostak is examining connections between telomerase, the aging process, and cancer.

Many researchers hope that our growing understanding of the processes of genetic replication and cell division will eventually lead to ways to slow and even reverse physical manifestations of aging, such as declines in strength and flexibility, lagging rates of healing, and memory loss. Through a special drug regimen or genetic engineering, perhaps hundred-year-olds of the future will have the same level of vigor as thirty-year-olds today. Aged nuclear tycoons, for instance, could revitalize themselves and maintain their enterprises as long as they'd like, much to the delight of their personal assistants (if their assistants are anything like Smithers, that is).

Inevitably, with radical new technologies come troubling ethical dilemmas. For example, what if embryonic human cells could be engineered not only to eliminate dreaded genetic diseases, but also to alter cosmetic characteristics such as hair color and thickness, eye color, skin pigmentation, expected adult height, and so forth? Would parents strive to have designer children? Would many couples try to custom-order a child with the precocity of Lisa, rather than with the impertinence of Bart? Ay caramba!

Meanwhile, science is edging closer and closer to producing novel life forms in the lab. Like many technologies, these could be used for humanity's benefit, through remarkable new cures, or its decline, through devastating new bioweaponry. As Lisa experienced through the outcome of her science experiment, by engineering our environment and tinkering with the basics of life, our responsibility to guard nature grows with our abilities to create and destroy.

"The Genesis Tub" is not the only *Simpsons* segment to feature an extraordinarily rapid evolutionary process. The episode "Homerazzi" includes a clever introductory sequence depicting the evolution of Homer from a single-celled organism into his modern-day form. It starts with Homer-faced cells quickly dividing, crying out "D'oh!" each time they split. These primordial organisms evolve into various aquatic creatures, including a Homer-like amphibian that crawls onto the land. In short order, a Homer-like ape emerges from the jungle and transforms into several different Homer-like humans. When the modern Homer finally takes his place on the family couch, Marge admonishes him with, "What took you so long?"

Perhaps if Marge studied the fossil record she would be more patient. The ground beneath Springfield and numerous locales around the world contains unmistakable evidence that life developed over billions of years. Let's now delve into Springfield's sacred soil and see what relics turn up.

7

Look Homer-Ward, Angel

Springfield is a town that values its history. Tourists flock to "Olde Springfield Towne" to learn what life was like in the days of Jebediah Springfield, its founder. Yet to drink in the full story of Springfield, it is not enough just to visit a mock colonial settlement and sample historical Squishees. The buildings that rest on Springfield's soil, and the people (such as perpetually inebriated resident Barney Gumble) who stage mock reenactments of the past, are only part of the full picture. Beneath the town's surface are relics of an older geological prehistory, fossil evidence representing the eons before human occupation. Thus, if you think some of Springfield's residents are troglodytes, dig a little deeper for the real thing.

The center for Springfield's prehistoric heritage is a field called Saber Tooth Meadow, from which many fossils have been brought to museums. Saber-toothed cats, or smilodons, roamed the prairies and woodlands of North and South America during the last Ice Age, until they went extinct some 11,500 years ago. They had prominent upper canine teeth and powerful bodies the size of modern African lions. With short legs they probably couldn't run very fast, but would sneak up on their prey. Skeletal remains of these large cats have been found throughout the midwestern region of the United States. The name of Springfield's fossil site, therefore, probably indicates that such bones were found beneath its soil.

As a promising junior scientist and environmental activist, Lisa feels very strongly that Springfield's prehistory should be preserved. Therefore, she is horrified when, in the episode "Lisa the Skeptic," she learns that Saber Tooth Meadow will be bulldozed and paved over to become the site of a major new shopping mall. What if there are undiscovered relics there that would be lost to archeology? Incensed by the failure of the mall developers to permit archeological excavation, she hires an attorney, the bumbling Lionel Hutz. Despite Hutz's incompetence, Lisa still manages to gain a promise from the developers (after some suspicious secret chatting among them) to allow some digging.

To find diggers, Lisa cashes in on a favor Principal Skinner owes her and gets him to lend her some student help. After a full day of digging, with assistance from Jimbo, Dolph, Kearney, Ralph, and others, Lisa stumbles upon a buried skeleton. Removing it carefully from the ground, the team examines it and finds that it looks remarkably like a skeletal angel with a full set of wings. Many onlookers, such as Ned and Moe, immediately conclude that the skeleton offers proof that biblical angels really exist. Flabbergasted, Lisa racks her brain for a sensible scientific explanation (such as the remains of a mutant) but can't think of anything reasonable.

Homer, entrepreneur that he is, hauls the skeleton to his garage, puts it on display, and charges admission for people to see it. Pilgrims flock to the "angel" hoping that praying to it will cure them of their maladies. Meanwhile, Lisa scrapes off a bone sample and brings it to the Springfield Museum of Natural History, hoping that a scientist on staff will be able to identify it using DNA analysis or other means.

That museum must have been exceptionally well funded at the time, because its resident expert is none other than the leading paleontologist and well-known author Stephen Jay Gould. Gould, who spent most of his career at Harvard, was codeveloper along with Niles Eldredge of the punctuated equilibrium theory of evolution, an alternative to the more widely accepted "gradualist" viewpoint. Briefly, the difference between punctuated equilibrium and gradualism is that the former proposes that evolution took place in fits and starts, with rapid growth spurts (induced, perhaps, by sudden environmental changes) separated by long intervals in which little happened, while the latter posits a more or less continuous record of evolutionary steps. This distinction is sometimes phrased as "evolution by jerks" versus "evolution by creeps." In addition to his vociferous contributions toward this debate, Gould also made his mark as a historian of science, wrote a regular column in *Natural History* magazine, and published his phonebook-sized magnum opus, *The Structure of Evolutionary Theory*, shortly before he died of cancer in 2002.

Gould was a well-respected but controversial figure for positioning himself as an ardent Darwinist with a nontraditional view. Thus, ironically, he needed to defend his position against both Darwinian purists as well as opponents of evolutionary theory, such as creationists. This became something of a Ping-Pong game with Gould's critiques of gradualism volleyed back by some of evolution's opponents as evidence of its "insurmountable flaws" and need for it to be supplanted (or at least supplemented) by an approach literally based on biblical accounts. Thus, appearing on an episode about an evolution controversy was not much of a stretch for him.

Although it was not surprising for Gould to guest star on the show, his role was unexpectedly anticlimactic given his great scientific stature. After Lisa gives him the bone scraping, he promises to analyze it. Later he runs up to Lisa, seemingly in a hurry to tell her something. When she asks him for a report he replies simply,

"Inconclusive," and then excuses himself. He ultimately reveals that he never even bothered to do the testing. Lacking expert results, Lisa has lost her opportunity to defend scientific methods and now must face her critics empty-handed.

Indeed, it seems that the whole town has ganged up against Lisa and science. News anchor Kent Brockman mocks Lisa for her steadfast belief that the skeleton is not an angel. Commenting on the need for mystery in life Ned says, "Science is like a blabbermouth who ruins a movie by telling you how it ends." After Agnes Skinner rallies the crowd to smash up scientific institutions, a group rushes toward the Museum of Natural History and begins to destroy dinosaur skeletons and other artifacts. It's one of the darkest days for reason since the time of the Spanish Inquisition. Who could have expected it?

At that point, events take a turn toward the truly bizarre. The "angel" mysteriously disappears from the Simpsons' garage and, after a search, turns up high on a hill overlooking Springfield. Lisa and others head for the hill and notice that etched on the angel's base is an ominous message announcing that "the end" will transpire at sunset. Reverend Lovejoy proclaims that Judgment Day is at hand.

Sundown arrives, and the residents of Springfield gather on the hill. As the sun sinks beneath the horizon, the people brace themselves for their doom. Seconds later, the angel starts to speak and rise off the hill. Now even Lisa, despite trying to rationalize what is happening, appears genuinely scared, clutching tightly onto Marge's hand for support. The angel begins to move toward the new shopping mall and announces its grand opening. Instead of doomsday, though, "the end" to which the angel was referring turns out to be the end of high prices. Lisa realizes that her discovery has been a shameless publicity stunt all along. Thus, like Piltdown Man, the "Springfield angel" has turned out to be just a clever hoax.

Because it was one of Gould's final media appearances, the episode became the focus of much commentary, discussed in an eclectic array of journals ranging from *Science and Spirit* to *Socialism*

Today. In the latter magazine, Pete Mason wrote, "Gould would have been very happy with this [episode] as his obituary. His references to popular culture (particularly baseball) are a defining mark of his essays, which appeared every month in *Natural History* magazine for nearly thirty years."[1]

William Dembski, one of the leaders of the intelligent design movement (the belief that life's complexity implies a designer), expressed a different view of Gould's performance. He remarked, "Gould comes off quite badly in the episode. Indeed, I'm surprised he let himself be used this way. To be sure, the religious fanatics and the simpleton townsfolk come off worse. But neither science nor religion triumph. Rather, it's *consumerism* writ large that emerges as the clear winner."[2]

Gould was certainly not the first advocate of evolutionary theory to step into a controversy. From the publication of his classic texts *On the Origin of Species by Means of Natural Selection* in 1859 and *The Descent of Man and Selection in Relation to Sex* in 1871, Darwin himself drew both praise and criticism. With his theory based on gradual change over time, it required a far older Earth than many of his contemporaries were prepared to accept. Moreover, by offering the then radical proposition that humankind was an animal species, it offended certain religious and moral sensibilities. Indeed, because of the controversy he realized his theories would create, Darwin delayed publishing his research for two decades, until his hand was forced by the independent discovery of evolution by another British scientist, Alfred Russel Wallace. When Wallace informed Darwin about his work, Darwin was astonished and decided to go ahead with publication. Despite Wallace's codiscovery, the theory has come to be known as Darwinism.

Both Darwin and Wallace were influenced by the dire theories of the Reverend Thomas Malthus, who predicted in 1798 that population growth would eventually overtake food supply, leading to an ever-increasing struggle for survival. Human population, Malthus argued, tended to grow at a geometric rate (continuously doubling), while food supplies could increase only at a much slower arithmetic (additive) pace. Therefore, eventually too many people would chase too few goods, and there would be large-scale famine. This could well lead to massive conflict, through which presumably only the strongest would survive. Picture the lunchtime struggles Homer would have with Lenny and Carl over the last available donut and magnify that by billions.

Malthusian ideas have turned out to be more applicable to animal and plant populations than to humans. Over the centuries since Malthus made his predictions, our species has developed increasingly advanced agricultural techniques, outpacing even its rapid population growth. Scarcity and hunger tend to be more the result of unequal distribution than the lack of enough food to feed the globe. Other species, on the other hand, obviously cannot increase their food supply through agricultural planning or importing from other places. Therefore, in regions with inadequate resources, they must compete with others to survive. In evolutionary theory, such a struggle is called survival of the fittest, a term coined by the British philosopher Herbert Spencer.

By making extensive studies of variations among animals and plants, Darwin realized the way that competition could lead to the introduction of new species over time. He collected samples and kept detailed journals during an epic voyage around the world on board the ship the H.M.S. *Beagle*. The five-year journey began in Plymouth, England, in December 1831 and included stops in the Canary Islands, all along the coast of South America, Australia, New Zealand, and South Africa. One of the highlights of the voyage was a detailed survey of the flora and fauna of the Galapagos archipelago, which Darwin found to be almost a world unto itself. There he encountered the famous giant tortoises, huge marine iguanas, more than a dozen species of finches, and numerous other exotic creatures. In his journal entries he noted the magnificent mosaic of varied characteristics the animals and plants possessed, such as the beak differences among finches. He came to realize that these varieties represented branches of family trees stemming back to common ancestors and began to posit that all living things were interconnected in such a fashion. Each variation, he reasoned, would present its own strengths and weaknesses in the struggle for survival, and would flourish or perish depending on how its features measured up to its competitors.

In trying to map out the entire network of how creatures on Earth were linked together by shared ancestry, Darwin sought fossils and other evidence that would provide proof of gradual transformation over time. Fossils generally result from the mineralization of an organism's remains while embedded in sedimentary material (such as a streambed) and offer a glimpse as to what its structure looked like when it was alive. In collecting and arranging these, Darwin came to realize that there were many gaps in the fossil record where no transitional species were evident. Gould and other advocates of punctuated equilibrium have pointed to these breaks as evidence for rapid change that occurs at sporadic intervals. For Darwin, however, the gaps seemed to result from the limits of the field of paleontology, as exhibited in the "poorness of paleontological collections." As he wrote in *On the Origin of Species*:

> Now let us turn to our richest geological museums, and what a paltry display we behold! That our collections are imperfect is admitted by every one . . . many fossil species are known and named from single and often broken specimens, or from a few specimens collected on some one spot. Only a small portion of the surface of the earth has been geologically explored. . . . No organism wholly soft can be preserved. Shells and bones decay and disappear when left on the bottom of the sea, where sediment is not accumulating.[3]

The publication of *The Descent of Man* motivated an intensive search for the "missing link" connecting humankind and apes with their common ancestors. Hope of fulfilling this quest clouded the vision of those taken in by the Piltdown Man hoax. Today, radiometric (using the percentage of certain radioactive substances to determine age) and other modern dating techniques have made such frauds increasingly unlikely. These methods have established that life on Earth dates back at least four billion years. They could have been used on Lisa's "angel" skeleton to establish whether all the bones came from the same individual. If they didn't, it would have been clear that the finding was a fake.

Dating techniques have been a boon for evolutionary theory, because they have shown that there has been sufficient time for random variations combined with the pressure of natural selection to lead to the full range of natural species. The brilliance of Darwinism is that chance, selection, and the passage of time work in tandem to produce species well suited for their habitats.

Chance is a curious thing. It's been said that if enough monkeys were pounding on typewriters for a sufficiently long time, they would reproduce the works of Shakespeare. That's because the monkeys would in due course type every possible combination of letters. So even if it took billions of years, they would eventually replicate anything Shakespeare or any other writer once wrote.

What about Homer puttering around in his garage? If he spent enough time messing around with a box of spare parts—trying to assemble them into various combinations—would a fantastic new invention eventually emerge? With sufficient diligence, could he even replicate the feats of Edison?

PART TWO

Mechanical Plots

Lisa, in this house, we obey the laws of thermodynamics!

—Homer Simpson, "The PTA Disbands"

Why can't I tinker with the fabric of existence?

—Lisa Simpson, Treehouse of Horror XIV

8

D'ohs ex Machina

What makes a mechanical genius? Why was Thomas Edison such a brilliant success? Though Edison himself claimed that genius was a mixture of inspiration and perspiration, that was clearly before the advent of contemporary models of learning and modern air conditioning. Today researchers have proposed a variety of theories about the nature of exceptional intelligence. For example, certain theories assert that there is a correlation between brilliance and social deficits.

Some think of Homer Simpson as just a dim bulb, and therefore would rule out any connection between him and Edison. However, there are commonalities shouting out as loud as the first phonograph. As reportedly in Edison's case, Homer seems to have issues dealing with people. Might that be a sign that tinkering with machines is his true calling? Could untapped brilliance lie beneath a veneer of utter incompetence? Might Homer's glazed look, like the glaze on donuts, cover the scrumptious and savory insights lying within? Like the famous inventor, Homer has been known to stare at glowing tubes—in Edison's case, experimental prototypes for incandescent light sources, and in Homer's, football games on television—but still the general concept is the same: electrons yielding their energies and illuminating glass. Both figures have strong ties to the energy industry. While Con Edison, the corporate successor to Edison's original enterprise, used to run nuclear power plants, so does Homer, sort of. Well, it's a stretch, but let's consider what happens when Homer tries to be truly inventive.

In "The Wizard of Evergreen Terrace," a play on Edison's famous moniker "The Wizard of Menlo Park," along with a reference to the Simpsons' street address, Homer aspires to be an even greater inventor than Edison. Motivating him is a pressing fear that he's wasted his life and doesn't have any accomplishments for which he'll someday be remembered. This restlessness begins when he hears on the radio that the average life expectancy is 76.2 years—exactly double his own age (at least what he thinks it is)—meaning that statistically his life is half over. Sulking, he ruminates that his time on Earth is halfway done with nothing much to show for it.

After Homer mopes around the house for some time, his family throws him a surprise party in an attempt to cheer him up. They show him movies of his accomplishments on an old-fashioned movie projector. After watching a few scenes from his life, Homer is disappointed when the film starts to burn from the heat of the projector. Upset with whoever invented "stupid movies," Homer is informed by Lisa that it was Edison. She also tells him about Edison's many other inventions, including the light bulb, the microphone, and the phonograph. Not believing her at first, Homer visits the elementary school library (because he's banned from the public library) and reads a number of children's books about Edison's life and achievements. Soon Homer has a new role model.

Homer's attempt to emulate and surpass "the Wizard" inspires him to become an inventor himself. At first he seems to lack the spark. He visits Professor Frink, who advises him to think of things that people need but don't yet exist, or, alternatively, novel uses for existing products. Homer's first thought is earmuffs made out of hamburgers. Frink seems to dismiss these, but he's actually already invented them.

Homer returns home, cloisters himself for a while, and manages to develop several new inventions, which he presents to his family. It's been an arduous process, so he hopes that they will applaud what he has developed. Instead they are baffled by the uselessness of the four products he unveils. The first is an electric hammer that pounds automatically but is hard to control and ends up bashing holes in the wall. Next is something that looks like an emergency alarm or smoke detector, called an "Everything's O.K." alarm, that beeps continuously only when nothing is wrong. It makes an extremely annoying repetitive sound and can't be turned off. Fortunately for the family's ears and nerves, it breaks right away. The third is a "makeup gun" that looks like a rifle and covers Marge with a smear of colors. Last is a "toilet chair," a living-room armchair that doubles as a working toilet.

When Marge frankly points out the worthlessness of Homer's inventions, he's gravely disappointed. Musing over his failures, he leans backward in a chair that he's specially adapted. The chair has a third pair of hinged, flexible legs that swing back to prevent it from toppling over. Marge and Lisa marvel at the cleverness of Homer's construction. He's ecstatic that he's finally stumbled upon something useful and unique. Crowing before a poster of Edison that he's tacked up in his basement, Homer looks closely and realizes that the inventor is sitting on a virtually identical chair. Edison made such a chair already, yet for some reason never marketed it.

In a fit of desperation, Homer decides to drive to the Edison museum and smash the original chair so he can continue to claim it as his own invention. With Bart in tow, Homer sneaks off from the museum tour, takes out his electric hammer, and is about to destroy the rival chair when he notices a poster on the wall. It compares Edison's inventive progress to that of the renowned Italian Renaissance thinker Leonardo da Vinci. Homer realizes that Edison was just as jealous of da Vinci as Homer is of Edison. Feeling sympathetic toward Edison, Homer decides not to smash the chair. Inadvertently, he leaves the electric hammer behind, which is found by the museum staff and announced on the news as a previously unknown example of another great Edison invention. The episode closes with Homer dismayed that he didn't even get credit for his own invention, a device that will probably earn millions for the Edison estate.

The museum in this episode is based on a real location where Edison once worked. Situated in the town of West Orange, nestled in the sprawling suburbs of northern New Jersey, the Edison National Historic Site stands as a monument to Edison's inventive genius. Its museum of inventions is housed in a complex of oddly shaped and intricately connected redbrick buildings that were once Edison's extensive laboratories. Museum visitors cannot help but be amazed by the numerous array of shelves stacked with roughly sketched diagrams, the hundreds of glass cases filled with rudimentary light bulbs and electrical gadgets, as well as the many rows of mechanical contrivances hanging from almost every ceiling, connected to strange machines on the floors. In those cluttered rooms, Edison used to toil for days at a time—getting almost no sleep at all except for brief naps—until he had worked out solutions to his technical problems.

One of Edison's inventions that offered him particular pride was the phonograph (Greek for "sound writing instrument"), also known as the "talking machine." He was the first man in history to record and play back the human voice. The pivotal events in the phonograph's development, exemplifying Edison's inventive process in general, took place in the summer of 1877. Edison was working on an instrument to transcribe the dots and dashes of a telegraph message onto a piece of paper tape for storage. Part of his device, helping to keep the tape in proper adjustment, was a small steel spring. Edison noticed, to his surprise, that as the tape passed along the spring a barely audible but distinctive sound was produced, resembling a human voice. Being highly innovative, he was inspired by this display to devise a mechanical means of recording sound by using imprints on similarly impressionable materials. By moving a stylus over such indented substances, he concluded that recognizable tones, including those of human voices, could be reproduced.

Edison soon drew up a working design for the first phonograph and sent it to one of his trusted mechanics. The machine that he developed and that the mechanic constructed included a tinfoil-covered metal cylinder imbedded with a fine spiral groove, a large screw mount upon which the cylinder would turn, a handle to rotate the cylinder, and a recording needle that would ride along the outside of the cylinder, following the spiral pattern. The other end of the needle was attached to a diaphragm, similar to those used in telephones. Thus, by speaking into the diaphragm and turning the handle, he would induce the recording needle to vibrate, creating an imprint on the tin foil, while at the same time causing the cylinder to revolve, distributing the needle's impression over the entire length of the spiral. The result would be a sequence of "hill-and-dale" indentations spread out over the cylinder. To play back the message, he would simply place the cylinder in a similar mechanism, but with a reproducing needle (stylus) and diaphragm instead of the recording apparatus. By turning the crank and listening to the sounds emitted by the diaphragm as the needle moved over the bumps on the tin foil, he could hear a more or less exact reproduction of the original message. The induced vibrations of the membrane would simulate the reverberations of human vocal cords and create realistic-sounding ersatz voices.

After his assistant, following careful guidelines, constructed a working model of the phonograph, Edison decided to put it to the test. Preparing the cylinder for recording by covering it with a fresh coating of tin foil, he spoke loudly and clearly into the diaphragm while cranking the handle. The words he uttered were hardly profound: "Mary . . . had . . . a . . . . little . . . lamb." The diaphragm vibrated, the needle bobbed, the cylinder spun, and the message was recorded. Then came the moment of truth. Edison replaced the recorder with the reproducer, turned the handle once more, and listened carefully to the sounds produced by the spinning cylinder. Sure enough, out came the exact words that he had spoken, reproduced in his own tone of voice. Although he had anticipated that this would happen, he was truly astonished by the results of his creation. He was amazed that he was actually hearing himself speak, minutes after the fact. The phonograph would become one of Edison's great commercial successes.

In another parlor of the museum, visitors can see where Edison set up a primitive motion picture camera for early experiments in cinematography. The impetus for Edison's invention may have been a suggestion by Eadweard Muybridge, inventor of the zoopraxiscope (a multiple camera system for capturing motion) that his device be combined with the phonograph to record both sight and sound. Instead, Edison decided to create his own motion pictures using a single camera, and set out to invent the kinetoscope (Greek for "watching movement"). Much of the work on the first movie camera was done by one of Edison's assistants, the photographer William Dickson. The original design was similar to early phonographs, with film attached to a turning cylinder. Later, when long, flexible celluloid sheets became available, this was replaced by spooled film.

On the grounds of his laboratory, in a cramped, dark building called the Black Maria, Edison established the first motion picture studio. There he recruited various kinds of performers (jugglers and so forth) to come and be filmed. Though the movies were short and simple, they heralded a revolution in how moving images could be recorded and established the United States as a major center for filmmaking and the entertainment industry in general. Thus we can thank Edison for the technology that paved the way for *The Wizard of Oz*, *Casablanca*, *Citizen Kane*, and *The Simpsons Movie*.

The light bulb, the phonograph, the movie camera, the filmmaking industry, and so much more—what *didn't* Edison invent?

For one thing, he didn't invent the electric hammer. A U.S. patent for a device by that name was issued to Hiroki Ikuta in 2005. Also, Edison didn't invent the wheel or even reinvent the wheel. As Carl points out in the episode, it was the Scottish engineer James Watt who developed the steam engine, one of the key innovations that powered the industrial revolution.

One lesson that this episode teaches us is that being an accomplished inventor does not require just extraordinary talent but also the right mix of problem-solving abilities to tackle the thorny puzzles awaiting him or her. To make his or her mark on history, a brilliant thinker must be at the right place at the right time. No two geniuses are alike, and no two situations alike, so there's a certain amount of luck needed for circumstances to be suitable. Edison's dogged persistence and ability to brainstorm creative solutions proved perfect for applying electrical laws and mechanical principles to the industrial and household needs of his time.

If Edison is one of Homer's scientific heroes, Einstein is certainly one of Lisa's. In Treehouse of Horror XVI, where a witch transforms townspeople into their Halloween costumes, Lisa ends up as Einstein. She even imitates his manner of speaking while figuring out how to break the spell. It is fitting that Lisa would identify with a thinker known for his theoretical insights and humanitarianism, while Homer would admire a more pragmatic man seeking business success.

Not that Einstein didn't also have a practical side. When he completed his university studies at the Swiss Federal Institute of Technology in Zurich, he tended to gravitate toward laboratory situations and skip classes with abstract mathematics. Later he would find that some of the very classes he skipped would be necessary to his research. Not surprisingly, he did not get good reports from his instructors and found difficulty, at first, getting an academic job after completing his degree.

Fortunately, Einstein had a friend with connections who helped land him a position at the Swiss Patent Office in Bern. It turned out to be a very rewarding and productive period in his life, offering him balance between time to perform his own theoretical calculations and the opportunity to earn a good income while engaging in a vital task. As patent officer, Einstein's role was to look over the blueprints and specifications for new inventions to assess if they were novel, useful, and feasible.

As Homer's example shows, not everything an inventor's mind concocts represents a startling new discovery. Not everyone can be as original as Edison. An item that lands on a patent officer's desk could represent something already patented but little known. Even if the item is original, it could be as useless as an Everything's O.K. alarm. Finally, it could seem a terrific idea but prove physically impossible to function. If it violates the known laws of physics, that's a sure sign it couldn't work.

Even the greatest inventors, Edison included, could not create a machine able to run forever. A savvy patent officer such as Einstein would unquestionably reject any "perpetual motion" scheme that he came across. Perpetual motion would violate the laws of thermodynamics, a key component of physical theory. Who would have thought that the inventing business would be so complicated?

9

Perpetual Commotion

Television cartoons may be the closest we'll ever come to perpetual motion devices. Although each week the characters might immerse themselves in harrowing situations, risking life and limb, by the time the next episode arrives, for most cartoons everything reverts to its original state. Even death, in cartoons, is perpetually reversible.

Take, for example, Bart and Lisa's favorite cartoon, *The Itchy and Scratchy Show*, a cat and mouse story with a twist and a lot of blood. It's quite possibly the most violent children's cartoon ever made. It may have even driven up the price of red ink. Itchy is the heartless mouse who devotes every waking moment to torturing the poor feline Scratchy. Scratchy has been vivisected, electrocuted, chopped up, and immersed in acid, and that's just for starters. In one segment, Itchy wraps Scratchy's tongue around a rocket that blasts off to the moon, bringing Earth's satellite crashing down on him. The amazing thing is that despite all this trauma, by the next episode Scratchy is always back for more. Nine lives are nothing for that cat. He can be reconstituted quicker than instant soup.

In real life, however, there is much that is not reversible. Drop a tray of china plates on the floor and watch them shatter into a thousand pieces, and they're not going to be featured at your next dinner party. Any type of explosion generally can't be reversed. Plunking ice cubes into a hot beverage is a good way of cooling it down, but you're not going to get those ice cubes back once you've finished the drink. These systems are irreversible due to deep physical principles—called the laws of thermodynamics—governing how energy transforms itself and how heat flows between objects of different temperature.

Cartoons, on the other hand, are notoriously immune to the precepts of physics. They don't need to respond to the pull of gravity, the buoyancy of water, or the power of the wind. Instead they approximate the laws of nature—or not—depending on whether animators are seeking realism or the bizarre. In 1980 a piece by the humorist Mark O'Donnell in *Esquire* magazine attempted to codify the physical laws of cartoons. Titled "The Laws of Cartoon Motion," it included principles such as "Any body suspended in space will remain in space until made aware of its situation," "Certain bodies can pass through a solid wall painted to resemble tunnel entrances; others cannot," "For every vengeance, there is an equal and opposite revengeance," and, what is undoubtedly Scratchy's favorite, "Any violent rearrangement of feline matter is impermanent."[1]

Given their habitual flouting of ordinary rules, it's certainly not a stretch for cartoons to feature perpetual-motion machines. Lisa creates such a device in the episode "The PTA Disbands." The motivation for her invention is that a lengthy school strike had made her bored out of her mind. The lack of class work and homework simply drives her mad. She is obsessively asking everyone around to grade her, assess her, pass judgment on her, reward her with A's and praise, and so forth, until Marge and Homer are driven bonker's as well.

Lisa has taken to developing her own science projects at home, which would seem on the surface to be a good thing. Yet her parents don't approve. Marge is upset that Lisa has been cutting up her own raincoat as if it were a dissection project. What rattles Homer is a device Lisa has rigged up that keeps whirling faster and faster. Somehow it is acquiring more and more energy out of thin air. As would any concerned father, Homer decides to take action. Lisa's obsessions must stop. Homer musters up the words of wisdom to set Lisa back on track, informing her that while she's at home she must "obey the laws of thermodynamics."

The strike eventually ends. Even Bart, who has been upset about the strange parade of substitute teachers, including his mom, is happy to be back at school. Lisa, obedient daughter that she is, follows Homer's instructions. Nary a word is heard again about her perpetual-motion device.

Ordering a real person to follow the laws of thermodynamics, the law of gravity, or any other physical principles is, of course, absurd. Our bodies automatically comply with the inherent precepts of physical reality. If any of these could be violated, then they wouldn't be laws.

That said, physicists sometimes don't know the proper arena within which certain laws apply. The laws of thermodynamics are an excellent example. Although it's clear that they apply to all things observed (not just to a single house, as Homer implied), physicists cannot say for sure if they encompass the *entire* universe. That's because they're applicable specifically to closed systems (in which matter and energy do not enter or exit), and it's unclear if the cosmos as a whole can be so characterized.

Let's examine the laws of thermodynamics and consider how they rule out perpetual-motion machines. Historically, these laws were discovered in the nineteenth century as a reaction to the development of the steam engine. Physicists such as Sadi Carnot, Rudolf Clausius, and William Thomson (Lord Kelvin) examined the question of what kinds of engines and processes could transform temperature differences into usable work, and developed what became codified as four distinct principles. We shall consider these in the order they came to be classified rather than in the sequence they were originally formulated.

The zeroth law of thermodynamics helps us define the concept of temperature. Defining temperature wouldn't seem to be such a tall order; after all, weather forecasters refer to it all the time, and we seem to understand what they are talking about. We owe that clarity to the consistency of working thermometers, which operate on the basis of law zero. It was numbered zero, by the way, because the other three laws were enumerated first, but the zeroth law seemed even more fundamental.

Thermometers work though a process called thermal equilibrium. If two things are in contact—one hot and the other cold—the hotter object transfers energy to the colder until both objects reach the thermal equilibrium state. The energy exchanged is called heat. When two bodies are in thermal equilibrium, heat no longer passes between them, and we say that they have the same temperature.

Now here's where the zeroth law comes in. Suppose you have two beakers of water. You place a thermometer in the first and wait until it settles. After writing down the temperature it reads, you shake it and position it in the second beaker. The thermometer settles down again, and you record the temperature of that beaker, too. If indeed the two temperature readings are exactly equal, then you can make a solid prediction. You don't need to be Einstein to guess that the two beakers of water, if placed in contact, would be precisely in thermal equilibrium and would not exchange any heat. That's because, according to the zeroth law, if two systems are each in thermal equilibrium with a third (the thermometer), they must be in thermal equilibrium with each other.

The next two laws are the meatiest, and historically the ones formulated earliest. The first law of thermodynamics is also known as the law of the conservation of energy, namely, that energy cannot be created or destroyed, but merely transferred from one variety to another. Einstein's addendum to this, expressed in his famous equation $E = mc^2$, is that mass is just another type of energy.

In nature, energy comes in many different forms. Heat is only one of its guises. Another type, called kinetic energy, is associated with an object's motion. The faster something moves, the greater its kinetic energy. A glass of water, for example, contains a colossal number of water molecules, each of which is moving. Therefore, it has a certain quantity of kinetic energy. For systems with an enormous number of components, we designate temperature as a measure of the average amount of kinetic energy per molecule. The hotter something is, the higher its temperature, the greater its amount of kinetic energy per particle, and the faster, on average, its molecules are moving.

Consequently, one application of the first law is that heat transferred to a substance can increase the average kinetic energy of its molecules, resulting in a temperature increase. That's why parents such as Marge and Homer would be wise to keep infants such as Maggie away from scalding sources of heat, such as radiators on full blast or the cores of nuclear reactors. For Homer, this precaution should be strictly followed even on "Bring Your Daughter to Work" day.

Speaking of Homer's job, another mode of energy transfer is work. Not only is work something he must occasionally do, it also has a technical meaning in physics. It's when force (or pressure) is applied to move something, such as pushing down on a button. If Homer is just sitting in place, he's technically not working, but once he lifts up a finger, plops it down on a button, and causes the button to compress, that fits the bill.

According to what is called the work-energy theorem, the application of work can cause a change in kinetic energy. If Marge pushes Maggie's stroller, for instance, she is applying work to it and thereby making it move faster. Consequently, Marge's work has transformed into kinetic energy.

Work can also alter an object's potential energy, the energy of position. If you lift something high into the air, your work increases its potential energy—a form that is recouped if you let it go and it falls. So if Homer was sitting far away from his computer console, in the midst of lunch, and an alarm rang indicating an emergency, he could perform his job indirectly, thanks to the wonders of potential energy. He could hurl a cup upward, transforming the work of his pitching arm into potential energy. When the cup lands, its potential energy would further transform into the work of pressing down the button that releases the water to lower the core temperature of the nuclear plant. Cool—potential energy is truly our friend.

Homer appears to have a certain ethical stance against expending extra energy. Hence his dismay upon viewing Lisa's perpetual-motion machine, which draws the energy needed for it to speed up faster and faster seemingly from out of the blue. Such a device would clearly violate the first law, since it does not conserve energy, but rather manufactures more and more from nowhere.

Could a device conserve energy and run with 100 percent efficiency (that is, zero waste)? According to the second law of thermodynamics, such a situation would also be impossible. Physicists express the second law in several different forms. One way of stating it sets a maximum limit to a closed system's efficiency that always falls short of 100 percent. That means that devices receiving no energy from the outside world eventually wear down.

For a steam engine, which works by converting some of the thermal energy of steam into the work driving a piston or a turbine, the maximum theoretical efficiency is set by the temperature difference between steam and the atmosphere. The second law mandates that efficiency rises with temperature difference. Thus, no engine could extract the considerable thermal content of the world's oceans unless it expelled a portion of the oceans' thermal energy into an even colder reservoir.

Another way of expressing the second law involves the concept of entropy, or disorder. Entropy is a measure of the lack of uniqueness of a physical system. Over time a closed system tends to progress from unique, orderly states to more common disorderly arrangements—hence either maintaining or increasing, but never decreasing, its total entropy.

Take, for example, a box of checkers sorted into piles of red and black pieces. You place the red checkers on one side of a checkerboard and the black checkers on the other side. With its uniquely divided, well-organized setup, the set of checkers is in a state of relatively low entropy. Now suppose you tap the board. At best, the checkers would maintain their separation, but they could also mix together. If you keep tapping the board, and the checkers mix more and more, the system would likely become less and less orderly and also less and less unique. Therefore its entropy would increase.

If you filmed the checkers mixing together and played the footage on a video player, first forward and then backward, you would see the checkers mixing together in the first case and separating themselves in the second. By watching the footage, you could readily tell the difference between increasing and decreasing entropy. Because the first situation would seem much more normal than the second, even without looking at the video controls you would know whether the film was running forward or backward. Thus the direction of increasing entropy provides a natural "arrow" of time.

Another application of the law of entropy involves objects of different temperatures brought into contact. For instance, suppose a double sink is filled so that one side has hot water and the other has cold water, and then a stopper that divides the two is removed. Before the stopper is pulled, when the hot and cold water are separate, the system is more organized than afterward when the temperatures start to equalize and the system is mixed up. Hence the system moves *naturally* toward greater entropy and lesser order, rather than the other way round. You could *artificially* reverse the process by heating up one of the sides, but that would render it an open system rather than a closed one.

Suppose Lisa wants to design a perpetual-motion machine that simply runs forever at the same rate, rather than one that goes faster and faster. She uses a battery to heat up a flask of water, which turns it into steam, and then uses its pressure to run a turbine. The turbine powers a generator that recharges the battery. The battery, in turn, reheats the water, and so forth. Because total energy is conserved, the device wouldn't violate the first law. Nevertheless, by perfectly recycling *usable* energy and having no waste, it would violate the second. Realistically, each time around the cycle, the law of entropy would ensure that not all the energy from the steam could be utilized—some would need to be released as excess. In other words, the system couldn't run at 100 percent efficiency and could never supply enough power on its own to recharge itself. In general, because of the second law, no machine could power itself completely through the temperature differences it created itself.

Finally, we come to the third law, pertaining to the impossibility of reaching the absolute zero of temperature. Absolute zero, a temperature of –273.15 degrees Celsius or –459.67 degrees Fahrenheit, corresponds to a state of no molecular motion whatsoever. Although scientists have cooled substances to ultra-low temperatures approaching absolute zero, physical principles guarantee that they could never remove all of a material's thermal energy. If you could run a heat engine such that its output was channeled into a reservoir cooled to absolute zero, it would run with 100 percent efficiency. However, the fact that absolute zero would be impossible to obtain is another reason why 100 percent efficiency would similarly be out of the question.

In short, the laws of thermodynamics guarantee that perfectly efficient devices would be impossible to create. Perpetual motion, though a fabulous idea for baffling science projects and curious conversation pieces, simply cannot transpire in our world of conserved energy and accumulating entropy. So if you ever receive e-mails advertising machines that run forever, you can safely delete them.

All this talk of efficiency is tiring. In this stressful age, there comes a time in everyone's life when they could use some help with completing their responsibilities. Homer, for example, could certainly use some assistance—be it from man or machine, or perhaps even a hybrid of the two.

10

Dude, I'm an Android

Let's talk about work, or better yet, about getting out of it. As we've discussed, according to the first law of thermodynamics work and energy are conserved, so if you don't want to do it, you've got to find somebody else who will. Someone needs to be trained to do all the things you are being paid and respected for, so that you can get the goods, the glory, and the deep family love without having to lift a finger.

They say that you can't teach an old dog new tricks. Therefore it would be hard to burden a pooch like Santa's Little Helper with extra responsibilities, as tempting as that sounds. But could you teach new tricks to robots? If new tricks don't match well with old flesh, could they be taught to gleaming machinery with sparkling and efficient movable parts? Perhaps. Yet another option would be to saddle kids with the chores, given that youthful human brains are vastly more impressionable and trainable than even the smartest mechanical minds. That line of reasoning follows the adage "Silly robots, tricks are for kids."

Thanks to the wonders of technology, someday we might not even have to choose. What about combining the exhilaration of youth with the obedience of robots? Robot children—legally constrained to remain at home until they are no longer minors, yet programmed never to talk back—could prove quite handy for getting little jobs done like sweeping the floor, taking out the trash, or constructing needed additions to the house. After all, android kids would never tire, remaining strong and peppy until they are switched off at night.

In "B.I.: Bartificial Intelligence," a Treehouse of Horror XVI segment parodying the film *A.I.: Artificial Intelligence*, the stork of modern technology drops a robot boy into the Simpson household, and they experience life firsthand with a junior android. Just why did they want a robot boy? The tale begins as a tragedy. In a fit of bravado, Bart—who, though flesh and blood, seems here to have a screw loose—tries to jump backward from an apartment building window into a swimming pool, crashes to the ground, and ends up in a deep coma. He remains in a hospital bed, completely unresponsive. Dr. Hibbert informs a devastated Marge and Homer that Bart will likely never recover and counsels them about a company that produces humanlike mechanical replacements. They purchase a "Robo-Tot" named David—visually indistinguishable from a human child, but with the advantage of durable parts and an instruction manual.

David quickly integrates himself into the family and makes himself indispensable. Marge is impressed by his cheerful helpfulness in the kitchen, the garden, and elsewhere around the house. Maggie enjoys getting a teddy bear her "brother" produces. Any ethical reservations Lisa has are alleviated by a friendly neck rub.

Then a miracle occurs: Bart awakens from his coma. Returning home and seeing his replacement in action, he naturally becomes extremely jealous. Anything Bart can do, David seems to do better. When Bart gives Marge flowers, David rigs up flashing electric building signs with loving messages. Bart simply cannot compete with David's circuitry, which is programmed to please parents.

Ultimately, Homer decides that there's room for only one of the two, and drops Bart off in the middle of nowhere. Wandering through the fields, Bart encounters a colony of defective androids, who beg him to teach them the meaning of love. Instead, Bart steals some of their parts and transforms himself into a mechanical powerhouse. He returns to Evergreen Terrace and destroys David (slicing through Homer in the process).

Parental love is a powerful bond. Instinctively, mothers and fathers cherish their offspring, by and large. Could they come to feel the same affection for a mechanical substitute? Would an android boy or girl call forth similar nurturing tendencies and offer a comparable level of emotional satisfaction?

A long-running study at the Massachusetts Institute of Technology Media Lab, called the Sociable Robots project, is designed to explore the emotional relationship between humans and robots. The project currently centers on two highly expressive mechanical beings: Kismet, an "infant" face with reactive, changing features, and Leonardo, a fanciful gremlinlike creature with long floppy ears. Overseeing the project is the robotics professor Cynthia Breazeal, who has pioneered an innovative hybrid between artificial intelligence and social psychology.

Kismet, the first of the two to be developed, began its life in the 1990s as Breazeal's doctoral thesis project. Its simulated facial features—including eyes, eyebrows, lips, ears, and so forth—are highly mobile, enabling it to mimic a considerable range of human expressions. People observing Kismet are usually able to sense what emotion it is trying to convey, as well as which direction it is gazing. Therefore it is able to look at someone and smile or sulk depending on its "mood." Kismet can also utter various childlike sounds to express the depth of its "feelings." If David the android is comparable to Bart, at least in terms of simulated age, Kismet is more like Maggie.

Studying and replicating social interactions is a two-way street, and Kismet is well equipped for observing people, too. Mounted on the upper part of its face are four charge-coupled device (CCD) electronic cameras. Two of these, constrained to move along with the head, have wide-angle vision and are used to gauge distances and take in the entire field of view. The other two cameras, located right behind the pupils, can move more independently and focus on closer objects. Depending on what Kismet is attending to, it can adjust the direction of its gaze.

For hearing, Kismet relies on wireless microphones and speech recognition software. Such software is similar to the automated voice-activated response systems that sometimes request information on the phone (when a recorded airline voice asks you to say "window," "aisle," or "wing," for instance, to indicate your seating preference). Those interacting with Kismet speak directly into the microphones. Their voice signals are transmitted to computers, where they are translated into instructions Kismet can understand.

Kismet's learning algorithms analyze the audiovisual input, combine it with other sensory input, and use the complete data to decide what it should do next. It could, for instance, turn its head, redirect its gaze, change its emotional state by altering its facial expression, or utter a response. The idea is for it to learn through imitation and experience how to socialize with humans. By watching Kismet unravel the fabric of interactive behavior, scientists may begin to understand the nuances of social learning.

One of Kismet's limitations is that is has no body. If you mention that fact to it, it might start to pout, so if you ever meet it face to face, you might wish to keep that observation to yourself. Because some types of emotional expression are nonfacial, Kismet cannot deliver the full range of interactions.

To boost interactive robots to a higher level, Breazeal's group commissioned the Stan Winston Studio to build Leonardo.

Standing two and a half feet tall and with sixty-one independent ways it can move its face and body, Leonardo is one of the most expressive robots constructed to date. Its facial movements alone are almost as complex as the possible modes of human expression.

The Stan Winston Studio has had considerable experience with robot making, particularly for Hollywood movies. It designed the robotic teddy bear Teddy for the film *A.I.*, the dinosaurs for the *Jurassic Park* films, the *Terminator* androids, and numerous other creatures, costumes, and special effects. Thus, the studio was a natural choice for bringing Leonardo to life.

Meeting Leonardo is a bit like stumbling upon a storybook creature from an enchanted forest. Though it looks like no known animal, its outsized furry ears, puppy-dog eyes, and wee proportions seem calculated to convey the friendship and warmth a talking pet might bring. Supporting its highly animated appearance is a bevy of tiny but powerful motors, enabling refined motions equivalent to subtle human gestures. Observing how people interact with Leonardo, and how Leonardo responds, has provided insight into the steps needed to master social communication.

In 2003, David Hanson of the University of Texas, Dallas, unveiled a robotic head, called K-bot, with realistic-looking polymer "skin" and a complete range of humanlike facial expressions. Like Kismet and Leonardo, it has electronic cameras for its eyes and minute motors to shift its gaze and transform its appearance. Because its skin is made of a flexible material, shaped by twenty-four artificial muscles, its facial movements are even more refined. In less than a second, K-bot's visage can shift from frowning to smiling and even to smirking or squinting. It could herald a new generation of lifelike robotic faces.

Even if a robot's face is able to move just like a human's, however, that doesn't mean that no one can tell the two apart. The wrong expressions, such as smiling repeatedly during inopportune moments, are a clear giveaway. As Breazeal has remarked, designing lifelike androids "isn't just an engineering problem" and must take into account social and psychological factors.[1]

We are on the threshold of a new age of machines with humanlike appearance, designed to physically and psychologically mimic our gestures and movements. The first use of these lifelike robots, seen already in rudimentary form at theme parks and toy stores, will undoubtedly be as entertainment devices. It's fun to see and hear mechanical beings imitate human expressions and respond to our spoken words. A number of companies are already manufacturing humanoid robots that walk, gesticulate, speak with a limited vocabulary, and respond to voice commands.

When could kids like Lisa and Maggie look forward to meeting robotic brothers and kids like Bart dread being replaced? If the mechanical siblings just need to look like humans and imitate human gestures, then the wait won't be long at all. Progress in making lifelike faces and bodies is proceeding at an astronomical pace. If, on the other hand, kids would like realistic conversations with intelligent companions that seem just like other kids and think just like other kids, then the wait could be much, much longer. Nobody knows if a robot could be ever designed that would pass the Turing Test, the Olympic high bar for machine intelligence.

The Turing Test, proposed by the British mathematician and renowned codebreaker Alan Turing in his influential 1950 paper "Computing Machinery and Intelligence," involves the critical question "Can machines think?"[2] Turing proposed to resolve the issue through an experiment called the imitation game. As he described it, the game would involve a human interrogator in one room and two respondents in another: a person and a machine. The interrogator would type up questions and transmit them to the other room. Without revealing who or what was answering the question, one of the respondents would reply. Then it would be the interrogator's task to guess if the answer came from a person or a machine.

Whether or not the Turing Test would reveal true intelligence is controversial. In 1980, the philosopher John Searle of the University of California, Berkeley, advanced what is known as the Chinese Room argument against claims that computers passing the Turing

Test would really be thinking like a person. Searle envisioned a closed room in which workers who did not understand a word of Chinese would be given pages worth of instructions in that language. To reply to these instructions, they would meticulously consult a rulebook that listed the appropriate response for each. The workers would write these answers down without having the faintest clue as to what the words actually meant. Outside the room, those transmitting and receiving the messages might think that they are engaging in a conversation with fluent Chinese speakers. Yet the workers would be like automata, lacking any understanding of that language. Similarly a computational system that simulated the responses of a human mind, Searle argued, would not necessarily be able to cogitate itself.

Almost six decades after it was proposed, the Turing Test remains a hard standard to fulfill. Computers today can record enough possible moves and look far enough ahead to beat the world's grandmasters in chess. Yet no existing computer has the creativity, flexibility, colloquial understanding, and intuition required to pass the Turing Test. None even comes close—as one can see by perusing the transcripts of the Loebner Prize contest for artificial intelligence, a real-life Turing Test competition that has taken place annually since 1990.

Winners of the Loebner Prize each year are the programmers who are judged to have developed the most intelligent conversational software. Their programs are gauged as Turing would: with judges, computers, and human confederates. For each round, both the computer and the confederate claim to be real people, and it's the judges' job to guess which is which. To make this assessment, each judge tries to carry on a remote conversation with each person or computer and decides who can best keep up and offer reasonable responses.

The judges for the 2005 contest were three New York University professors and a journalist: Ned Block, Lila Davachi, Dennis Shasha, and John Sundman, respectively. The program that came in first place, written by Rollo Carpenter, is called Jabberwacky.

Strangely enough, its personality bears at least a passing resemblance to Nelson Muntz, the leading Springfield bully.

For example, when Ned Block asked it, "Which is further away from us, Queens or Inner Mongolia?" Jabberwacky responded, "Ha! Ha! That was stupid. Tell me a *good* joke this time."

When John Sundman told it, "I live on Martha's Vineyard," Jabberwacky replied, "I'm going to Mars and am going to smash your face with a baseball bat!"[3]

Hmmm. Maybe Lisa and Maggie need to wait a bit longer for a cuddly android brother they could play with. But if they need an android bully to taunt them, that's a different story!

What if the Simpson kids were ever on the market for a robot father? Suppose they needed a mechanical mentor who would make any sacrifice for their welfare, no matter how painful. Who could possibly fit the bill, except Homer Jay Simpson himself?

11

Rules for Robots

What if Homer woke up one day and suddenly became industrious and efficient? Suppose he always followed instructions to the letter and never slacked off before completing a task. Picture him strong and capable, working harder than any typical man, and winning the love and respect of his kids. Imagine Homer as a robot.

In the episode "I, (Annoyed Grunt)-bot" Homer makes just that transformation, albeit in a clunky kind of costume. The tale begins with Bart teased about his little bike and wanting a big flashy new one. After Homer informs him that he'll get a newer model only after his old bike is defunct, Bart conveniently has the old one wrecked by placing it in front of Dr. Hibbert's moving vehicle. Apologetic, Dr. Hibbert offers to pay for a new bike. Homer agrees to buy one, and off he and Bart go to the store. Assembly costs extra, so Homer decides to do it himself. Bad move. When the brand new bike falls apart because of Homer's incompetence, Bart is extremely peeved.

Homer decides to make it up to Bart and prove his mechanical prowess by entering a father-son robot competition on the popular show *Robot Rumble* (based on the television show *BattleBots*, hosted by the company of the same name). The idea behind the show is that fathers build fighting robots for their kids and bring them to an arena, and they brutally battle—putting their metal and their mettle to the test. The trouble is that despite Homer's hope to be another Edison, his mechanical skills make a door hinge seem dexterous. After his efforts to construct a mechanical gladiator prove futile, he recalls Abe's paternal advice, "If you can't build a robot, be a robot."

Slipping into a robotlike outfit, equipped with a bogus remote control and a mallet for fighting, Homer enters the competition himself—at first keeping his true identity secret from Bart and everyone else. In Homer's robot guise, Bart dubs Homer "Chief Knock-a-Homer." Homer, leaving behind a rather lame excuse note to explain why he couldn't attend the fight in person, plays a convincing robot fighter. Going up against a mechanical opponent wielding a buzz saw, aptly named Buzz Kill, he manages to suffer a slice to the arm without screaming and ultimately prevails. Bart is brimming with pride.

After Knock-a-Homer successfully battles several other robots, he faces his most formidable foe, a colossal fighter designed by Professor Frink. In the first round, Frink's robot clobbers him until he's dazed and confused. When Bart takes a look beneath Knock-a-Homer's back panel to see if there's any damage, he's startled to discover that Homer's been inside all along. Nevertheless, he's still very proud of his dad—perhaps even more so now that he realizes Homer's sacrifice.

There's not much time for family bonding, though, before Knock-a-Homer is literally dragged into round two. Frink's colossus bashes Homer so hard that the robot suit opens up and he's squeezed out like toothpaste. Then instantly the fighting stops.

Frink explains that his robot follows Asimov's Three Laws of Robotics and cannot injure a human being. On the contrary, it is programmed to serve humans. Demonstrating its devotion to *Homo sapiens* (in this case to "Homer sapiens"), it mixes up a martini and plunks Homer down on a comfy lounge chair. Ah, that's the life.

The episode raises two vital questions about robots: What in the world are Asimov's Three Laws of Robotics, and could a robot mix a credible martini? Let's address the latter issue first. Strangely enough, since 1999 the city of Vienna has hosted Roboexotica, an annual cocktail-mixing exposition for those with a chip in their shoulder. Robotic "Moes" from around the globe demonstrate their bartending skills, much to the delight of Viennese "Barney Gumbles." The festivities include the Annual Robot Cocktail Awards for robots with skills in mixing and serving cocktails, lighting cigars or cigarettes, and/or making bartending conversation. Given the interplay at the Loebner competition, you can imagine what the casual banter might be like. Some of it is in fact deliberately insulting for the purpose of attracting attention. If a robot's not quite ready for the Turing Test, however, at least it might master the pouring test.

Could people someday have robot servants, programmed to make their masters' lives as comfortable as possible? If the robots were strong enough to do heavy lifting, they could indeed be dangerous. What if the equivalent of Jimbo, Dolph, and Kearney pooled their classmates' lunch money and used it to program an android to beat up every kid in sight? Or what if a notorious criminal like Snake Jailbird trained a mechanical titan to rob Kwik-E-Marts? Dangerous robots able to be programmed to commit heinous acts clearly wouldn't be acceptable. Extraordinary power would require extraordinary precautions.

The visionary writer Isaac Asimov pondered this situation in his famous robot stories, collected in the anthology *I, Robot*, on which this episode was loosely based. Asimov was a Russian-born,

American-raised trained biochemist who became an extremely accomplished writer of science fiction as well as fact. He was very concerned about the moral implications of new technologies, especially if they could be misused for violent purposes. To head off the possibility of malevolent dynamos bent on destruction, he proposed the Three Laws of Robotics. First featured in his 1942 story "Runaround," they have since passed into the annals of science fiction legend. The laws are:[1]

1. A robot may not injure a human being or, through inaction, allow a human being to come to harm.
2. A robot must obey orders given it by human beings except where such orders would conflict with the First Law.
3. A robot must protect its own existence as long as such protection does not conflict with the First or Second Law.

During *Robot Rumble*, Frink's mechanical combatant followed these laws to the letter. It protected itself until Frink gave it orders to engage in combat. Then it halted all aggressive activity once it was in danger of harming a human. Finally, it sensed that Homer needed to relax (a kind of implicit request) and offered him the martini and the lounge chair. Not bad for an early-twenty-first-century machine.

Be advised that at least for now Asimov's laws have only fictional applications. Unlike the laws of thermodynamics, they are purely hypothetical, since it would presently be impossible to program a robot to render ethical decisions. Yet what if someday androids do have the power to protect or harm, based on choices they make? Perhaps Asimov's laws will serve as a blueprint for built-in safeguards against misuse.

In advanced industrial nations, especially those with a low birth rate such as Japan, there is a realistic possibility in years to come that humanoid robots will become part of the workforce. In 2000, Honda introduced a sophisticated mobile robot called Asimo (Advanced Step in Innovative MObility) that looks like someone in a white helmeted spacesuit. Though its name seems a tribute to Asimov, apparently the similar appellation is pure coincidence. Nevertheless, with its ability to walk and run like an agile human being, Isaac Asimov's near-namesake represents a milestone along the path to the robotic society he envisioned.

Within decades, perhaps, robots could become a familiar fixture of daily life—sweeping floors, serving food, and assisting the homebound. If we start trusting our fate to machines, we would certainty insist on restrictions in their programming that would prohibit them from intentionally causing harm. That's where something like the Laws of Robotics would come into play.

Without such safeguards the world could become as treacherous and unruly as an Itchy and Scratchy cartoon. Who could predict when androids would turn against their creators, like Frankenstein's monster did? Perhaps only those experienced in chaos theory would have the mathematical tools needed to anticipate whether well-behaved robots could suddenly run amok.

12

Chaos in Cartoonland

In hundreds of episodes the Simpsons have experienced utter mayhem, complete pandemonium, absolute bedlam, and an unholy mess. They are well acquainted with anarchy, turmoil, discord, and confusion, and have caused many a ruckus. Catastrophe, calamity, havoc, and discombobulation seem to afflict them during the rare moments they aren't in the midst of total disaster. Homer alone has brewed many a brouhaha, and often these have led to many "d'ohs." Yet in only one episode do the Simpsons learn the true meaning of chaos.

The term *chaos* in physics has a technical meaning that distinguishes it from the jumbled mishaps of life. This definition separates it from inexplicable, sheer chance happenings, and places it in a strange hybrid category between the worlds of perfect predictability and total randomness. Chaos implies that a system has underlying laws that theoretically permit knowledge of the future, but prohibit these advanced looks in practice because of uncertainties in measurement that grow over time. In other words, a scientist like Professor Frink might be able to write the equations describing a system (a chemical mixture, for instance), but unless he had perfect experimental devices that measured every aspect of it with absolute precision, he could not say for sure how the system would develop.

The quintessential application of chaos theory—and historically the initial impetus for the field—is analyzing and forecasting the weather, through the science of meteorology. Along with military uses, weather forecasting was one of the primary applications of the first computers, built in the midtwentieth century. Predicting whether it will rain, snow, sleet, hail, or just leave us alone requires the analysis of enormous quantities of data; thus in essence computers made weather prediction much more feasible.

Early computers were enormous compared to today's devices and far slower. Programming them initially required rewiring, then flipping switches. By the 1960s, punch cards came into wide use as formats for entering programs and data. A deck of cards, containing the steps needed to process the information, as well as the data itself, would be fed into the computer, which would run through all the steps and send the results to a printer, which would type them up on long paper sheets. Naturally, running through this lengthy process left plenty of room for error. A small error in typing up a punch card could completely change the outcome of a program, yet it might take hours or days to detect by running the program again and again and looking through all the cards.

Edward Lorenz was a respected MIT meteorologist adept with early computers and knowledgeable about the critical components of a forecast, including temperature, air pressure, and wind velocity. In 1960, he constructed a basic set of equations relating these quantities and using the laws of physics to predict their future values. Classical physics, developed by the seventeenth-century

English scientist Sir Isaac Newton, and furthered by the eighteenth-century French mathematician Pierre Laplace and many others, is completely deterministic. Determinism means that if you knew all the conditions of a system at one particular point in time, you could perfectly anticipate indefinitely into the future how these conditions would develop.

Take for instance, the game of pool. If a learned professor like Frink were playing the game and wanted to use the cue ball to sink the eight ball into a corner pocket, he could use protractors and rulers to measure various angles and distances related to the balls, walls, corner, and cue stick. He could use the laws of physics to determine at which speed and angle the eight ball should be hit by the cue ball to maximize its chances of hitting one of the walls, bouncing off, and landing in the pocket. Then he could practice aiming the stick at exactly the right angle and hitting the ball to produce the right velocity. Hence classical physics' deterministic equations would allow him to plan what will happen in the game.

Lorenz fully expected that the equations for the weather would behave in similar fashion. Theoretically, he thought, if enough data points were entered into a computer, it could plot out how the wind would shift, temperatures and air pressures rise and fall, and so forth, for each location in a region. Thus he confidently entered his series of equations and set of data into a computer and waited for the printout, hoping that it would at least approximate weather conditions.

To make sure his program worked Lorenz ran it twice with what he believed was the same set of data. Plotting out the results each time, he was astonished to find two different forecasts, looking increasingly dissimilar over time. How could the same information, plugged into the identical computer program, yield such disparate pictures?

Then, as a further check, Lorenz sifted through the numbers very carefully. As it turned out there was a very slight discrepancy between what he had entered each time. In one case he had truncated the data differently from the other, keeping a different number of digits. That's like writing Homer's age as 38.1 in one case and just 38 in the other. With such a slight difference, you would think the forecast wouldn't change much. If Homer told an actuary he was 38 and got a report than he would likely live 45 more years, then revised his estimate to 38.1 and was told that he would likely live only 10 more years, such an enormous difference would be very surprising. But for Lorenz's algorithm any tiny discrepancy cascaded over time and ended up being quite a major change.

Scientists have dubbed the phenomenon of small changes in the initial conditions leading to giant differences in future dynamics the butterfly effect. The expression arises from the possibility that the flapping of a butterfly's wings in the skies over one part of the globe could produce minute changes in air patterns that would escalate into major differences in the weather of another region. (Originally Lorenz put this in terms of a seagull's wings but would later give a talk titled "Predictability: Does the Flap of a Butterfly's Wings in Brazil Set Off a Tornado in Texas?"; hence the expression.) Because data are never 100 percent precise, Lorenz realized that the butterfly effect implied that weather forecasting has significant limits.

In 1963, Lorenz reported his findings in a paper titled "Deterministic Nonperiodic Flow," published in the *Journal of Atmospheric Sciences*. For more than a decade, because it appeared in a specialized journal, his article was little read by the physics community. Eventually, however, physicists outside of the field of meteorology noted its curious conclusion that deterministic equations could yield essentially unpredictable results.

A turning point was a 1975 article by the University of Maryland mathematicians James A. Yorke and Tien-Yien Li showing that the transition to chaos is a universal mathematical phenomenon for certain types of periodic deterministic systems. Something could operate in regular cycles under a range of conditions yet become effectively unpredictable if those conditions were slightly altered. This realization led to numerous experiments showing that chaotic behavior occurs throughout the natural world, from the fluctuating pace of water dripping from a faucet to the intricate arrangement of Saturn's rings.

The concept of deterministic chaos entered the popular arena through several different vehicles, including a widely read book, *Chaos*, by the science reporter James Gleick, and compelling articles on the subject in magazines such as *Scientific American* and *New Scientist*. However, it was Jeff Goldblum's nerdy mathematician character in the blockbuster film *Jurassic Park*—the "chaotician" Ian Malcolm—that would make chaos theory a synonym for supposedly well-planned science gone awry. In that film, based on the best-selling novel by Michael Crichton, biologists use intact dinosaur DNA to clone modern examples of those thunderous beasts, and then let them roam in a kind of theme park. Electric fences ensure that these powerful creatures are well confined. Despite the precautions, Malcolm warns that instabilities could blossom into unexpected behavior. Indeed, when some paleontologists and children visit the park, everything that could go wrong does, including a complete shutdown of the electric system when they are among the carnivorous varieties. Thus unfortunately Malcolm's "chaos theory" turns out to be right on the mark.

Jurassic Park was not the first disaster movie set in a theme park. Years earlier, the film *Westworld*, also written by Crichton, had a similar premise, only with lifelike robots instead of dinosaurs. The androids in *Westworld* brought realism to an extensive fantasy kingdom consisting of three themed lands: medieval, Roman, and Wild West. The Western-themed part (which gives the film its name) includes cowboy types with genuine weapons. When their computer systems malfunction, the robots, led by a mechanical marksman called Gunslinger, begin to attack the park guests. The film came out in 1973, two years before Yorke and Li introduced the scientific definition of chaos and well before that expression became popular. Consequently, though *Westworld* conveyed a strong message about the limits of supposedly predictable systems, no one in the film used the expression "chaos theory."

The Simpsons episode "Itchy and Scratchy Land" is a clever composite of *Westworld* with *Jurassic Park*, with the world's goriest children's cartoon thrown into the mix. Bart and Lisa beg their parents to take them on a family vacation to the most violent theme park imaginable. At first Marge and Homer say no. As they continue to refuse, Bart and Lisa badger them relentlessly and wear down their resistance. After learning about an adult-oriented Parents' Island on the grounds of the park, where grown-ups can chill while the kids get their thrill, Marge and Homer finally agree to go.

Attractions at the park include evading giant axes, skipping through mine fields, and tumbling down ultra-steep log flumes as sharp buzz saws slice through them. It's the sort of place where Vlad the Impaler would have sent his little ones for summer vacation. Not only are there rides galore, there's interactive entertainment as well. This takes the form of costumed versions of Itchy and Scratchy roaming through the grounds and giant robotic renditions marching in regular parades.

To replicate the "playful banter" of the cartoon series, each Itchy robot comes equipped with search and destroy capability targeted at locating and obliterating Scratchy robots. As a safety precaution, their digital cameras are linked to processors able to distinguish images of Scratchy robots from those of human beings. Thanks to this protective circuitry, they obey a version of Asimov's Laws of Robotics and are precluded from harming people.

After an exhausting romp through the park with the kids, Marge and Homer are ready to adjourn to Parents' Island. They bop to 1970s disco beats in a nostalgic dance club, while the kids enjoy more rides and shows. Unsupervised, Bart cannot help getting into trouble. When he can't resist aiming his slingshot at a costumed Itchy, he's whisked into an underground detention center deep beneath the park grounds. There he reunites with Homer, who has been detained for trying to kick another costumed Itchy. An embarrassed Marge has to come down and ask for their release.

The underground level of the park is buzzing with activity, the staff trying to maximize the enjoyment of the visitors above. A special unit repairs the robots that have been damaged in the parades. Somehow, Professor Frink is a witness to this operation and offers the repair staff a chilling warning. Using "elementary chaos theory" (as he puts it) he predicts that the robots will "go berserk" and turn against humans, but not for twenty-four hours. Frink is right about the revolt, but a bit off in his calculations—the robots begin to rampage almost immediately. The devices preventing them from harming humans seem to malfunction, causing the robots to stalk and attack the park's employees and guests.

Back on the park grounds, Homer doesn't realize at first why a mechanical Itchy is approaching him. He foolishly thinks the giant robotic mouse wants to be his friend. But then the malfunctioning machine goes on the attack, along with numerous other robots, and the Simpsons try to flee. A promising escape route eludes them when the park's helicopters fly off without them. Just in the nick of time, Bart realizes that flash photography confuses the robots' circuitry and makes them shut down. The robots collapse one by one, and the family is saved. The lesson seems to be that even the best laid plans of mechanical mice and men often go awry.

Could chaos theory enable researchers to predict catastrophe? Curiously enough, there's a related branch of mathematics called catastrophe theory, developed by the French topologist René Thom in the 1960s, that bears on such predictions. Catastrophe theory shows that a quantity could change slowly and continuously for a time, then suddenly jump or fall to a completely different value like taking a leap from a precipice. For example, the stock market could rise during a period in which stocks are overvalued, based on false hopes that profits will keep pace, only to take a rapid tumble when those expectations evaporate. Thanks to the research of the British mathematician Christopher Zeeman and others, catastrophe theory has even been applied to animal behavior, attempting to explain why dogs might glare at an intruder for a considerable time before starting to bark ferociously, as if some hidden threshold had been reached.

Chaos theory similarly posits that small changes in a quantity could produce massive changes to a system, converting it from as regular as a ticking clock to as unpredictable as a dice roll. Though chaos mimics randomness, research has revealed common milestones along the road to disarray. In the mid-1970s, the mathematical biologist Robert May showed that a simple equation, called the logistic map and indicating how the population of a species develops over time, possesses a kind of tuning dial that transforms its dynamics from stable to periodic and ultimately to highly erratic. If a parameter representing a species' reproductive rate is cranked up beyond a certain value, the species might begin to produce more offspring than its environment can sustain. Its next generation might be smaller because of lack of resources, causing the population to dwindle below its ideal size. With the population decreased below its supportable amount, the succeeding generation could grow bigger again, and so forth. This rhythmic effect is known as a population cycle with a period of 2.

Increase the reproduction rate parameter a bit more, and the population starts to cycle between four distinct values, a dynamic change known as bifurcation or period doubling. Nudge the parameter up again, and a cycle of period 8 commences. In each case, the population grows and diminishes in a regular fashion that returns to each level after a finite number of steps.

If the parameter is cranked high enough, a strange thing happens. No longer is there a semblance of regularity. Rather, the population level becomes as sporadic as the results of a roulette wheel. No random factors have entered the equation; it's still governed by the same deterministic formula. Yet chaos has emerged from regularity, like a multihued bouquet pulled out of a staid black hat.

Around the time that May published his seminal work on this topic, the American physicist Mitchell Feigenbaum used an early programmable calculator to make an independent and surprising discovery about the path to chaos. Experimenting with an equation similar to the logistic map, Feigenbaum measured the rate of progression of period doubling and found that it converged on a special value: approximately 4.669. Then he took a number of completely different equations, calculated how quickly period doubling progressed for each of those, and was astonished to discover that each homed in on the same constant. Today that value, a novel mathematical constant unrelated to any other, is called the Feigenbaum number. Its existence demonstrates that amid the transition to pure chaos there is considerable order.

Once full-blown chaos ensues, it comes draped on a framework of regular patterns. The keen eye (or computer program) can spot these regularities and use them to make detailed predictions. For example, the numerical results of Lorenz's meteorological equations, if modeled through computer graphics, curiously resemble a butterflylike formation. Strangely enough, any point in space that is not on the butterfly's wings, if entered into the equations, ends up gravitating toward one of the wings. Conversely, two nearby points on one of the wings, if plugged in to the formula, tend to move onto separate wings. It's like a crowded, high-priced resort that tourists strive to get into, but once inside, they try to get away from one another as much as possible. Chaos researchers call such a mixture of drawing in and spreading out a "strange attractor."

Strange attractors possess an intriguing mathematical property called "self-similarity," meaning that any patch, if magnified, resembles the whole thing. Self-similarity abounds in nature, from the twigged branches of a tree looking somewhat like the tree as a whole to the sinuous banks of a stream resembling the shores of a much greater river. In 1975, the French mathematician Benoit Mandelbrot coined the term *fractals* to describe such self-similar structures, because their number of dimensions appeared to be fractional (instead of the one dimension of a line, two of a plane, or three of space).

Since the time of May, Feigenbaum, and Mandelbrot, researchers have applied chaos theory and the concept of strange attractors to a vast range of natural systems, hoping to use the orderly features embedded within the chaotic dynamics to render accurate forecasts. For example, the Harvard Medical School professor Ary Goldberger has used chaos theory and fractals for more than two decades to study the behavior of the heart and other aspects of human physiology. Using a mathematical analysis of electrocardiographic (ECG) results, he has proposed ways of understanding various types of heart arrhythmia. Some of his most recent work has applied fractal measures to the question of how aging occurs and disease progresses.

If chaos theory can be applied to the complex mechanisms of the human body, it could certainly be used to analyze robot behavior. As programmed mechanical systems, even advanced robots would exhibit deterministic behavior. A mathematical analysis of a robot's patterns of action could well reveal underlying loops and patterns, and in some cases sequences that appear random. Because this seemingly random behavior would flow from mechanistic internal directives, it would represent a type of deterministic chaos and could be examined through the techniques of that field. Hence it would not be a stretch for someone familiar with chaos theory (as Frink claims to be) to apply the methods of chaos in an attempt to anticipate whether robots might start acting in an erratic fashion.

Frink could similarly apply his chaotic forecasts to his own inventions, given that many of them end up causing havoc. He means well, no doubt, but sometimes he fails to take enough precautions to account for human incompetence. Take, for instance, the time he sells a teleportation device to Homer that possesses the dangerous capability of combining radically different creatures into freaky hybrids. Frink does try to warn Homer about the possibility of catastrophe, but given his predictive savvy, perhaps he should have kept such a perilous contraption under wraps. At the very least, he could have helped the Simpsons debug it.

13

Fly in the Ointment

In entomological lingo, commonplace at professional meetings such as the International Congress of Dipterology, where winged, six-legged creatures are the buzz, you could say that Bart is pretty fly for a guy. Entomologists study insects, and those specializing in diptera are well familiar with flies, mosquitoes, gnats, and so forth. For a guy to be fly by such exacting scientific standards, he needs to have just the right genes. That is certainly the case when, in the Treehouse of Horror VIII segment "Fly vs. Fly," a device Professor Frink develops accidentally mixes Bart's genes with those of a housefly.

Homer purchases the said device at a kind of "flea market" in front of Frink's house. That should have tipped him off right away as to what kinds of critters might merge in it with human DNA. On the face of it, it's a type of two-booth matter transporter—like Superman's private changing room, only doubled. Hop in one booth and you instantly pop out of the other, like someone quickly and surreptitiously voting twice. Although its listed cost is a whopping $2, Homer bargains Frink down to 35¢. Sweet.

The matter transporter seems useful at first. Homer no longer has to walk up the stairs. He just positions one booth at the bottom, the other at the top, and, presto, instant transport. By placing a booth in front of the refrigerator, and another elsewhere in the house, he has immediate access to his cherished Duff beer.

Then Bart sneakily starts to experiment. Shoving both of the family pets, Santa's Little Helper and Snowball II, into the machine at the same time, they emerge as double-headed and double-tailed hybrids of dog and cat. That gives Bart a wicked idea: to try to become a superhero with the body and mind of a human and the swift wings of a fly. He jumps into the transporter along with a fly, and two horrific beings come out. One is a tiny buzzing insect with Bart's head and personality (call him "Bart-head Fly"), the other is Bart's mindless body topped with the giant head of a fly (call him "Fly-head Bart").

Bart-head Fly flitters around and seems to have some fun. Threatened by a spider (in a scene reminiscent of the classic movie *The Fly*, on which this episode was based), he laughingly evades it. But then seeing what has become of his human body, he becomes jealous and worried.

Fly-head Bart is a repulsive monstrosity, emitting awful noises from his hideous face. Yet the Simpsons decide to embrace him as a full member of the family. His human qualities have become supplanted by the desire to flap his arms and consume enormous quantities of sugar and syrup. No longer does he have human traits such as patience, empathy, and the love for quiet contemplation (which I'm sure were somewhere deep down before his ghastly transformation). Ah, Bart; ah, humanity!

Meanwhile Bart-head Fly decides to contact Lisa and let her know who her real brother is. Seeing his profile in her desk lamp,

Lisa coaxes him into her saxophone, where Bart's voice resonates and can be heard. When Fly-head Bart finds out about this, he becomes jealous, runs after Lisa, and tries to eat Bart-head Fly. Lisa opens up the door to their microwave oven just in the nick of time and propels both of them together back into the matter transporter. Within the device, Bart and the fly's genetic material separate and resume their normal states. Bart walks out of the machine looking just like his old self and apparently safe and sound. Everyone seems happy to see Bart except Homer, who is suddenly and inexplicably furious about Bart using his matter transporter.

Compared to the practical devices we've been discussing, including Edison's classic inventions, heat engines, and robots, the notion of energizing matter and transporting it across space is extremely hypothetical. In coming years, it is doubtful that we'll witness people traveling instantaneously between distant locales. Converting the myriad atoms of a human being to pure information, conveying that immense quantity of data from one place to another, and reconstructing the same person from new material would present formidable philosophical issues and technological challenges to say the least, if it were even possible. Who would volunteer to be pulverized if there was even the slightest risk that they couldn't be perfectly reconstituted? Elementary particles are a different story, however. They are far simpler and lighter than people, of course, and don't bring up the messy issues associated with consciousness, volition, consent forms, lawsuits, and so forth. Thus they make ideal test subjects for this purpose. Currently, many researchers are investigating the question of instantaneously transporting particle characteristics in a process called quantum teleportation.

From its inception in the 1920s, quantum physics has inspired controversy about its counterintuitive implications, particularly its description of random, instantaneous occurrences on the atomic and subatomic levels. While according to classical physics scientists can theoretically measure any feature of nature with absolute precision, quantum mechanics includes a built-in fuzziness. A key ingredient of quantum mechanics, Heisenberg's famous uncertainty principle, mandates that certain paired physical quantities, for example position and linear momentum (mass times velocity), are impossible to measure simultaneously and precisely. In other words, if researchers exactly determine a particle's position, they cannot precisely measure its momentum at the same time, and vice versa. The more they know about one quantity, the less they know about the other.

The standard quantum approach, known as the Copenhagen interpretation because of the city where Niels Bohr and Werner Heisenberg and their colleagues developed it, asserts that before a researcher measures a physical quantity, its quantum state, encapsulated in a mathematical object called the wave function, typically corresponds to a range of possibilities. A particle's wave function provides information about the potential values of its physical quantities, distributed according to their likelihood. (Technically, it is the wave function *squared* that yields the actual probability distribution.) Plotted with respect to position, momentum, or another quantity, the wave function offers insight into how each of these parameters can remain vague before measurement but home in on a particular result as soon as the measurement occurs. The transformation from a distribution of possible values into a single result, chosen randomly, is called wave function collapse.

The collapse of a wave function resembles what would happen to Mrs. Krabappel's grade distribution if suddenly all the members of her class disappeared except for one random pupil. Before the disappearance, the distribution would look like what statisticians call a bell curve, reflecting the wide range of student abilities in her class. Afterward, it would look like a spike, centered on the performance of the single pupil left. Clearly if Martin Prince were the only survivor, the peak of the graph would be near the high end of the grades, and if Bart remained it would be somewhere very different. Similarly, when a quantum wave function collapses, its distribution of the quantity being measured suddenly becomes a sharply defined, randomly located peak.

An important constraint is that because of the uncertainty principle, it would be impossible for a wave function to collapse into sharply defined position and momentum distributions simultaneously. If it has a spiky position distribution, it has a spread-out momentum distribution, and vice versa.

The idea of probabilistic quantum collapse was anathema to Einstein, who vehemently objected that the divine plan for the universe would certainly not include dice rolling. He was also troubled by quantum physics' nonlocality, which he called "spooky action at a distance." This manifested itself when two interacting particles were represented by a common wave function.

In a paper with Boris Podolsky and Nathan Rosen, Einstein presented what is commonly known as the EPR (Einstein, Podolsky, and Rosen) paradox. Their argument was designed to show that quantum physics is philosophically unsavory because it appears to permit instantaneous communication between widely separated particles. This contradicted the long-held idea that communication between objects must take a finite amount of time, limited by the speed of light. A variation of the EPR paradox developed by the physicist David Bohm conveys this quandary in simple form.

Electrons and other particles possess a quantum property called spin that relates to their behavior when placed in a magnetic field. An electron, for instance, has two spin states, "up" and "down." In an analogy easy to visualize but not exactly physically accurate, we can think of electrons as charged spinning tops. If these tops rotate counterclockwise, their rotational axis points up, and if they rotate clockwise, it points down. According to magnetic theory, the up-pointing top and the down-pointing top would have opposite arrangements of north and south magnetic poles and would therefore behave differently if a powerful external magnet were nearby. Although an electron is not really rotating about its axis like a top, it shares with rotating objects two different magnetic alignments. Researchers can observe the two distinct spin orientations of electrons by analyzing atomic spectral lines.

According to the exclusion principle proposed by Wolfgang

Pauli, two electrons in the same location cannot have exactly the same quantum state and therefore must have opposite spin states. If one is up, the other must be down, like Ralph and Clancy Wiggum sharing a seesaw. An electron pair therefore must be in a "spin singlet," which means a mixed spin state combining the two orientations. Physicists refer to such a linkage as "quantum entanglement." Which of the pair is up and which is down can be determined only through measurement—causing the wave function representing the mixed state to collapse into one of two possibilities (up-down or down-up, as the case may be).

Now imagine producing an electron spin singlet in the lab and physically separating the two particles by a great distance. Move one to Alaska and the other to Florida if you'd like. Until you measure their spin, you wouldn't know which is up and which is down. Now place one of these electrons in a spin detector. The wave function associated with the mixed quantum state would instantly collapse. If the spin detector for the measured electron read "up," the other one's wave function would immediately collapse into a pure state of spin down. No matter how far away they are, there would be no lag time between the measurement of one and the transformation of the other.

Einstein found the concept that information about a quantum state could instantaneously pass from one point in space to another point far away and cause such a transformation extremely troubling. He believed that it violated the principle that the speed of light is the upper limit for the rate of communication. Consequently, he searched in vain for a more fundamental theory to explain the behavior of electrons and other elementary particles. Defenders of quantum theory point out, however, that no matter or radiation would actually be exchanged between the entangled particles. The determination of their spin states would just reveal correlated properties. Hence communication would never actually exceed the speed of light.

It's as if a husband and wife share two credit cards—one gold and the other platinum—and each spouse randomly grabs one of them each time they travel. Suppose the husband flies to Alaska for a business meeting, and the wife heads out at the same time to Florida for another convention. If the wife removes the credit card she brought from her purse and it happens to be the platinum card, she would instantly realize that her husband must have taken the gold card. Yet though she would have immediate knowledge of her husband's choice, no one would say that they exchanged a faster-than-light signal between each other.

Until the early 1990s, virtually nobody believed that something like the EPR experiment could be used for teleportation of the kind described in science fiction. In 1993, however, a team led by the IBM researcher Charles Bennett proved that properties could be stripped completely from one particle and bestowed on another. The implication is that the information needed to replicate an object could be fully transferred, assuming that the original is completely robbed of its identity.

Since then there have been a number of experiments confirming that quantum teleportation is possible as long as the primary object's properties are wiped clean. The farthest-reaching teleportation scheme to date was performed in 2004 and involved transferring physical properties across the Danube River in Vienna. A team from the University of Vienna, including Rupert Ursin, Anton Zeilinger, and five others, set up two stations, one on either side of the river. One was called Alice and the other Bob, and they were linked by a fiber optic cable threaded through a sewer tunnel. They used the Alice station to teleport a complete set of information about a particular photon (light particle) to Bob, particularly its state of polarization.

Polarization pertains to the angular direction that the electric field component of a light wave oscillates through space. For instance, it could wiggle like a vertical jump rope, a horizontal jump rope, or somewhere in between. It is one of the characteristic features of a photon, like a fingerprint.

To complete the transfer of properties, several critical steps were required. First, both stations needed to share another entangled set of photons that acted as a kind of codebook. The Alice photon was combined with one of the entangled photons, and a joint measurement was made. On the basis of the result of that measurement, the Alice photon was wiped clean of its polarization state and a signal was sent to the other entangled photon on the Bob side. As soon as the Bob photon received the signal, it transformed into the exact polarization state that Alice used to have. The end result was that the Alice photon's characteristics were teleported to the other side of the Danube, and essentially the Bob photon became Alice.

If this could somehow be done with people, something like Frink's transporter could be perfected. Imagine if Homer was standing in a booth on one side of the Danube, and there was a booth on the other side full of the precise material ingredients needed to reproduce his body. Suppose beams of entangled photons were sent to each portal. One of these would combine with Homer and a detector would analyze all the atoms in his body. Upon this analysis Homer would become a bland pile of inert material, and a complex signal would be sent to the other side. The beam would combine with the material and the photons that were already on the other side and would reconstruct the exact state of Homer's body. Suddenly he would find himself on the other side. We might picture him grabbing a "Danube Duff" from the local brew house (served perhaps by a Viennese automated bartender from the Roboexotica convention) and sighing contentedly.

Teleporting a person sounds almost feasible until you think about the enormous quantity of atoms in the human body and the profoundly serious implications of destroying someone to create a replica. Pioneers in quantum teleportation have emphasized that the state of the art involves far simpler systems than actual bodies. Zeilinger, for example, has pointed out that the challenges involved in teleporting people would be astronomical:

> We are talking about quantum phenomena here. We have no idea how we could produce these with larger objects. And even if it was possible, the problems involved would be huge.

> Firstly: for physical reasons, the original has to be *completely isolated* from its environment for the transfer to work. There has to be a total vacuum for it to work. And it is a well-known fact that this is *not particularly healthy* for human beings. Secondly, you would take all the properties from a person and transfer them onto another. This means producing a being who no longer has any hair colour, no eye colour, nix. A man without qualities! This is not only unethical—it's so crazy that it's impossible to imagine.[1]

Quantum teleportation is far from the only conceivable means of instantaneous transport. An even farther-reaching means of instant relocation, at least according to observers watching it happen, would be the hypothetical ability to stop time itself. If a person could move while everything around them somehow remained frozen in place, they could stroll from one point to another in literally no time at all. Such a strategy would be particularly useful for kids who love causing havoc but never find enough moments in the day to carry out their pranks. Know any kids like that?

PART THREE

No Time to D'ohs

So our kids keep getting smarter. If we have another one, it could build a time machine which we could use to go back in time and not have any kids.

—Homer Simpson, "Smart and Smarter"

Foolish Earthling! Totally unprepared for the effects of time travel!

—Kang, "Time and Punishment"

14

Clockstopping

Bart never has enough waking hours to pull off all the stunts his diabolical mind concocts. There are only so many seconds in the day to deliver painful wedgies, tie his classmates' shoelaces together, paint humiliating slogans about Principal Skinner on schoolhouse walls, shame Mrs. Krabappel about her love affairs, break off the head of Lisa's "Malibu Stacy" doll, weave in and out of honking traffic on his skateboard, fool Homer into letting him play violent video games, and so forth. And that, for Bart, would be just a good morning's work.

Poor Milhouse just can't keep up with Bart's antics. He desperately wants to be cool, whatever it takes, even if it gets him into trouble. Yet he is clueless about what real trouble is all about, and, like a first-year judo student, needs to eye the master's moves closely. Slow on the uptake, Milhouse could well use slow-motion instant replays of Bart's stunts for practice.

In the real world, time is not forgiving. Opportunities pass in a flash, and if one doesn't seize them it's just too bad. A moment's hesitation could mean the difference between the undetected placement of simulated barf on a teacher's chair and a trip to Groundskeeper Willie's shack for after-school detention and bagpipe lessons.

In the Treehouse of Horror XIV segment "Stop the World, I Want to Goof Off," Bart and Milhouse discover an amazing panacea for their time-management issues. An old magazine ad leads them to purchase a stopwatch that has the ability to freeze time. By simply clicking a button on the watch, everything else in the world, save the person or people touching it, stops dead in its tracks until the button is pushed again.

Grasping the watch together, and clicking it off and on at opportune moments, the crafty miscreants launch a reign of absolute mayhem. Each frozen interval offers Bart ample time to rearrange the people and things around him in the most devious, embarrassing, and hilarious way possible. Finally Milhouse is able to keep up with his comrade in crime and savor the fine art of producing pandemonium. No shred of dignity is spared, as the Springfield townspeople discover the wording on signs rearranged into nonsensical messages, Principal Skinner's pants suddenly pulled down during an assembly program, and Mayor Quimby's clothes replaced in succession with a maid's uniform, a colonial costume, and other strange outfits.

When the mayor finds a way of locating the culprits through their footprints in a special "ultraviolet powder," the citizens take up arms to try and capture them. Krusty's sidekick Sideshow Mel is determined to kill them before their secret is revealed. Fleeing from the revolting residents, Bart and Milhouse click the watch off, then accidentally drop and break it. Instantly, all movement ceases in the entire world, save the frantic efforts of the boys. Only after they have finally reassembled the watch—piece by piece over the course of fifteen years—does time's flow resume again.

Could time really be frozen and thawed out—like Krusty Burger meat before it's lovingly served by capable teenage staff? If so, would we find the results appetizing, or would we be as revolted as the Springfield townspeople (until they finally revolted)? Could such a process, if ever developed, find better use than simply goofing off?

You may have experienced a less extreme version of time flowing at a different rate when you are having a blast while those around you are bored out of their skulls. For example, if you are listening to a live broadcast of your favorite band, hours might seem to whiz by in a musical haze. On the other hand, if your family members are staring at the clock wondering when you'll remove your headphones and join them for dinner, they might protest that they've been waiting an eternity.

Psychological time—the time of the mind—is well known to be extremely variable. Many factors influence whether time seems to speed up or slow down, including the amount and level of activities you are engaged in. Psychologists believe there is a connection between the complexity of what you are doing and your estimation of how long it takes.

Aging also affects the perception of time's passage. Children have a much more expansive view of time than adults do. For a child of Bart or Lisa's age, waiting a month for a birthday present might seem impossibly long. Yet when Abe Simpson recounts his heroic actions during World War II, he speaks as if they happened only yesterday. Though we can chalk up part of that to his severe memory loss, it's clear that his life clock operates at a vastly different pace from that of his grandchildren.

Mind-altering drugs such as hallucinogens represent another known influence on time perception, as someone with Otto's pharmaceutical predilections might attest. For example, the drug DMT, an element of a tea used in certain native Brazilian religious ceremonies, seems to put the brakes on time's flow and unite all moments into one. (Perhaps that's why Homer once enigmatically reported that he was the first non-Brazilian to travel through time.) About the effects of this psychedelic substance researcher Rick Strassman wrote, "Past, present, and future merge together into a timeless moment, the now of eternity. Time stops, inasmuch as it no longer 'passes.' There is existence, but it is not dependent upon time."[1]

Could a pharmaceutical agent be used to freeze people literally in place? No known drug stops people precisely where they are standing, keeps their bodies like statues, renders their memories blank, and then allows them later to resume all activities as if nothing had happened. True, there are drugs known to cause temporary paralysis of various parts of the body—even to stop the heart during certain types of bypass surgery by flooding it with potassium. Such steps are not taken lightly and incur many risks and possibly permanent damage. Naturally, they are performed under general anesthesia, which acts temporarily to "freeze time" for the mind as well as for the body. Those awakening from an anesthetic slumber often feel the disorientation of having had hours of complete nonawareness.

More commonly, we cannot help but experience altered temporal states virtually every night. Ordinary sleep offers a colossal leap across great chasms of darkness, spanning seven, eight, or more hours in a state of restful timelessness. Have you ever dozed off so quickly that you didn't even realize it and woken up hours later startled by the glaring sun of a new day? It's almost like someone has clicked off and then restarted your personal stopwatch.

Dreams—the flight entertainment of sleep's voyage—offer even grander excursions along time's manifold byways. In nocturnal reveries a dreamer might feel that days or even months have passed while they've been dreaming for only a few minutes. Dozing off in Mrs. Krabappel's class, for instance, Bart might imagine a whole lifetime for himself as Radioactive Man, vanquishing foe after foe, only to wake up through a kick by Nelson and find out that he just slept through the recess bell.

Powerful emotions, such as extreme fear or anxiety, can also seem to stop the clock. Parents realize this during emergency situations when they must act quickly and a rush of adrenaline allows them to do so. As a concerned mother, for example, if something ever happens to Bart, Lisa, or especially helpless little Maggie, Marge's heart races and she takes action immediately with almost superhuman powers—that is, as soon as she notices they are missing.

Dreams, drugs, emotional states, and so forth alter the pace of our body rhythms and the rate of our temporal perception. Scientists are able to test such differences by asking research subjects without access to timepieces to estimate the duration of particular time intervals, then comparing their estimates to the readings of accurate clocks. These clocks, in turn, are calibrated by making sure they tick at the same pace as the best terrestrial standard, currently measured by atomic transition rates.

In the seventeenth century, Isaac Newton proposed that earthly clocks, in theory, could be set to the tempo of a universal rhythm, what he called "absolute time." In this view, perfect clocks, traveling through any region of space, could keep pace with one another no matter what their speeds or circumstances. In the early twentieth century, however, Albert Einstein found that in order to resolve certain physical contradictions this absolute viewpoint needed to be abandoned in favor of a relative perspective. His advances led physics to embrace a more flexible view of time—not just of our personal experience of it, but also of its fundamental nature.

The clashing principles Einstein needed to reconcile were two very basic physical propositions. The first is that all motion at constant velocities is relative. We observe this effect when we are riding in a closed, steadily moving vehicle, such as a smooth, slowly rising elevator, and feel like we are not moving at all. Conversely, we also notice it when we're in a stopped vehicle—such as a train at a station—look outside to see another train pulling out, and think for a moment that we're moving ourselves. Our senses inform us—and Newtonian physics confirms—that we cannot feel the difference between perfectly straight, uniform motion and not moving at all. The only way to distinguish the two is to look for background clues, such as objects moving past. Manipulating such background images can trick the eye and present the illusion of motion. Thus, if Chief Wiggum's police car is parked on a movie set and he sees projected images of scenery rushing by in the opposite direction, he might be fooled into thinking he is really chasing a suspect.

Einstein realized that the concept of relative speeds seemed to contradict another established physical principle, that the speed of light in a vacuum appears the same for all observers. Descriptions of light developed by the British scientist James Clerk Maxwell and others mandated that its measured velocity must be independent of the relative speed of anyone doing the measuring. Thus if the aliens Kang and Kodos aim a giant laser beam at Earth, and a ship of sympathetic beings tries to outrace the beam and rescue our planet, their efforts would be to no avail. With Kang and Kodos cackling madly to each other in the background, the good guys would realize that no matter how fast they travel, the light would always seem to be retreating from them at exactly the same speed, and they'd never catch up.

To explain the behavior of light through the physics of motion, Einstein found that he needed to replace the Newtonian concept of absolute time with an observer-dependent definition. He proposed the idea of "time dilation" as a way of two observers traveling at different velocities still measuring the same speed of light. Briefly this states that the clocks of those in a vehicle moving close to light speed run slower according to those not in the vehicle—for example, from the perspective of stationary observers on Earth. Because speed is distance divided by time, if someone's clock is running slower they could travel greater and greater distances during these laggardly intervals and still not exceed the speed of light. Thus in the case of the friendly extraterrestrials trying to rescue Earth, though they keep cranking up their engines and moving closer and closer to our planet during the time intervals ticked by their ship's clocks, they still can't beat out the laser beam.

Time dilation is one ingredient of Einstein's special theory of relativity, proposed in 1905. Another is "length contraction," the notion that objects moving close to the speed of light appear from the stationary (that is, not moving along with the object) point of view to be compressed in the direction of motion. For instance, if Kang and Kodos are zooming back to Rigel 7 at near light speed, those standing on the remnants of Earth, assuming they had powerful enough telescopes, would see the ship and its fiendish occupants squashed like rotten tomatoes along their homeward path. A third aspect of Einstein's theory is the interchangeability of matter and energy, epitomized by the famous equation $E = mc^2$, and helping to furnish the power behind Burns's vast nuclear empire.

Now let's consider one admittedly far-fetched, but theoretically valid, way of using Einstein's time dilation effect to design a kind of watch that would seem to stop time (or, more properly, compress it relative to conventional time on Earth). For this experiment, Bart and Milhouse would need their own ultra-high-speed spaceship (perhaps borrowing one from Kang and Kodos), and a special watch able to operate it by remote control. Imagine that whenever Bart and Milhouse click the watch, the spaceship is programmed to whisk whoever is around them (Skinner, Quimby, and so forth) off into space at a velocity close to the speed of light. Remaining in Springfield, the boys could make mischief to their hearts' desire (rearranging letters on message boards, breaking into Skinner's house and pasting "kick me" signs onto the seats of his trousers, and so forth). When they were finished with all their pranks, they would click the watch again, and the spaceship would return. The passengers would be astonished to discover that their possessions had been mysteriously vandalized in an incredibly short interlude, according to their watches.

For all practical purposes, however, enacting such a time-bending scheme would be next to impossible. For the "time stoppage" to seem swift, the spaceship would need to load the passengers and accelerate from rest to near light speed in the space of seconds, corresponding to liftoff forces that would be deadly by any measure. If the vehicle enacted a more reasonable rate of acceleration, then there would be a long interval when it would be catching up to speed. That would nix the idea of "clicking time off and on" in the manner of minutes or hours, and replace it with differences noticeable in the span of months or years.

For example, if a spaceship loaded with Springfield inhabitants (except for Bart and Milhouse) accelerates continuously away from Earth at a tolerable rate of 1 *g* (the acceleration of free-falling bodies on Earth), it could attain a velocity close to light speed in close to a year. It could cruise at that rate for a number of days, then return to Earth while decelerating. The net effect would be that the townspeople would be missing from Springfield for about two years from the ship's vantage point, but longer than that from Earth's point of view. If they traveled sufficiently close to light speed during their cruising interval, they could, for instance, miss fifteen years of Earth time. Ultimately, in that case, the result would match at least one aspect of the "Stop the World . . ." episode. While Bart and Milhouse would have aged a full decade and a half, bringing them into the prime of young adulthood, Skinner, Quimby, and the others would be only two years older.

Special relativity is not the only Einsteinian theory that allows clocks to move at different rates. A decade after completing his first monumental theory of space and time, Einstein brought forth an even greater masterpiece, his general theory of relativity. While the special theory pertains to ultra-high speeds, the general theory relates to gravity. To model how gravity influences the paths of objects, it describes space and time in tandem as a kind of a flexible fabric called the space-time continuum that curves whenever it is weighed down with matter. The greater the mass in a region, the more this fabric curves, like a hammock stretched by the weight of heavier and heavier bodies. If Maggie was placed on a hammock, for example, the hammock would hardly bend at all, but if Homer sat on it drinking a Duff, it would curve much more, and if Comic Book Guy donned a Super-Skrull costume and jumped on it, it might even break. Alas, such is the flimsy nature of physical reality. Super-Skrull impersonators of exceptional girth are just not appreciated by the space-time continuum.

The "hammock" in which Earth resides is the solar system, occupied in the center by the most massive body in our region, the sun. The sun's mass distorts our region, causing objects in its vicinity to curve their paths through space. In a similar fashion, if Homer, while sitting on a hammock, dropped an empty Duff can onto its fabric, it would either roll toward him or roll around him depending on how it landed. So because of the sun's curving effect, Earth "rolls" in an elliptical orbit around the solar system instead of moving unhindered in a straight line through space.

One of the most revolutionary aspects of relativity is that space and time are closely interrelated. Whenever space bends, time stretches as well. That's why length contraction and time dilation go hand in hand. Therefore, near a very massive object such as a star, time intervals are lengthened compared to those of empty space.

Perhaps the densest objects in the universe are the collapsed relics of stars known as black holes. About these, Comic Book Guy is undoubtedly a leading expert, as he has dallied with many collapsed relics of stars at science-fiction conventions. Black holes have captured physicists' as well as science-fiction fans' imaginations because of their unusually strong gravity and other captivating features. If these captured physicists ever escape, they would likely report that the extreme bending of space-time's fabric due to black holes' enormous concentration of matter would lead to enormous differences between the way clocks would run near a black hole and on Earth.

Traveling close to a black hole is another way time could be slowed down or even stopped compared to ordinary terrestrial time. This represents an example of time dilation due to exceptionally strong gravitational forces rather than high speed. Let's imagine a scenario in which Burns decides to blast some of his workers off into space so that they can investigate nuclear technology under vacuum conditions. Smithers equips the spaceship with monitoring devices to make sure the blasted employees don't slack off. Unfortunately their ship heads into the region of a black hole. As the craft approaches the collapsed star, passengers Lenny and Carl, oblivious to the approaching danger, might decide to play a round of poker. Due to the warping effects of the nearby black hole, their personal clocks would begin to tick at a slower and slower rate compared to Earth time—not noticeable to them but only to outsiders. Keeping a close watch on their activities, Smithers would observe their poker moves seemingly becoming more and more lethargic. Upon hearing about this, Burns might grumble that not only are his workers slacking off, they appear to be slacking off in their slacking off.

Each black hole is girdled by a zone of no return—an event horizon—that corresponds to the boundary of the region inside which escape would be physically impossible. If Lenny, Carl, and their coworkers enter that zone, their clocks would stop completely relative to Earth time. In other words, an infinite amount of Earth seconds would pass for a single second to pass on the ship. Smithers would observe the ship to be frozen forever at the bleak precipice of the event horizon. When Burns learns of this, he might be jealous of their apparent immortality. His envy would be misplaced, however, given that the workers would still experience their own time passing at its usual rate, while their ship is stretched out and torn apart by deadly gravitational forces.

If, on the other hand, they manage to pull away from the black hole just before they cross the event horizon, they could eventually return to Earth. Upon their homecoming, they might find that they have aged much less than those they left behind. For example, they might be astonished to discover that Bart and Milhouse are no longer boys, but well into their twenties, and that Homer has effectively been retired for thirty years.

Clock stopping is not an easy trick. Unlike joy buzzers, decoder rings, or faux driver's licenses, you can't purchase time-stopping watches from just any old kids' magazine. Nevertheless, the variability of our perceptions of how quickly events pass and the stretchiness of Einsteinian relativity each allow for the possibility that a minute for one person is an hour, a day, or even fifteen years for others. Two decades, for some ten-year-old boys, could be insufficient time for all the mischief they want to engage in.

15

A Toast to the Past

It is commonplace to believe that the past is gone; it's history, toast. But toast has a way of popping up again and again—acquiring the golden hue of nostalgia, or the burnt, acrid taint of regret, depending on the setting. Some try to cover up the past with the sweet, creamy butter of wishful thinking. When this turns out to be just a false, oily substitute, many simply can't believe it's not butter. Any way you slice it, however, if you try to microwave your memories, they end up soggy and barely palatable.

All this would seem like a crusty old metaphor if it weren't for the curious circumstance that in the Treehouse of Horror V segment "Time and Punishment," Homer literally reaches the past by pressing down the lever on an ordinary toaster. It was a broken toaster, you see, and Homer, in trying to fix it, had turned it into a crude time machine. Plunging headlong through the eons, he arrives at the age of the dinosaurs. In an obvious parody of Ray Bradbury's classic tale "The Sound of Thunder," he discovers to his horror that any change to the distant past, no matter how minor, snowballs over time into substantial differences for the present to which he returns. His father, Homer recalls, had admonished him on his wedding day about the possibility of going back in time and altering history, and now the reason for this warning has become all too clear.

For instance, the first time Homer arrives at the past, he crushes an annoying mosquito. This paltry death sets off a long chain of events extending forward through time like dominoes toppled one by one. When the toaster's lever pops up, Homer returns to a ghastly present in which Ned Flanders is supreme dictator. Like Big Brother, everyone in Springfield must obey him without question or else undergo a lobotomy that completely eliminates free will. Bart, Lisa, and Marge have all accepted Ned's authority; will Homer be next?

Escaping from Ned's forces, Homer presses the toaster's lever down again and returns once more to the age of the dinosaurs. Vowing not to touch anything and tamper with time, he nevertheless accidentally sits on a fish that has just walked out of the water and squashes it. Thus he once again has disrupted the fragile chain of events that has led to the familiar present. When the toaster lever pops up, Homer returns to his own time but discovers that the rest of his family are giants. Thinking he is a kind of bug that happens to look like Homer, the gargantuan Bart and Lisa try to pound him with their fists. By pressing down the toaster's handle again, Homer escapes barely in time.

In Homer's third excursion to the days of the thunder lizards, he sneezes, setting off a chain reaction that topples one dinosaur after another. Returning to the present, he braces himself for whatever bizarre changes are in store for him. Initially, he is delighted that his house and family seem pretty much the same with a few exceptions: his sisters-in-law, Patty and Selma, just died, his household is wealthier, everyone is polite to him, and his family owns a Lexus. "Woo hoo!" he exclaims.

Then comes the shock to his system—the cruel plot twist that sends Homer screaming and wailing in utter disbelief. Objectively, it is really a small difference between the alternative reality he has created through his primordial sneeze and the world he used to call home. Yet for Homer, that distinction shakes the core of his existence like a tornado rattling his soul. No one in this godforsaken universe has ever heard of donuts! As Homer hurriedly pushes down the lever on his time-traveling toaster, an ironic downpour of donuts marks his departure. Apparently donuts are quite common, but they are just called "rain." But it's too late; he has already rolled fate's dice once more.

Eventually, after a number of false tries, Homer does find a universe that suits him. Donuts are plentiful and people do in fact eat them. The only minor issue is what they eat them with: everybody has a long, reptilian forked tongue that reaches out to food and slurps it up. Oh, well; that's good enough for Homer.

The notion of traveling through time has been an indelible part of culture at least as far back as H. G. Wells's epic novella *The Time Machine*, published in 1895. The protagonist of that tale, an inventor referred to only as the Time Traveller, explains in the opening pages that because space and time are flip sides of the same coin, the fact that you can move through the former means that it is at least theoretically possible to journey through the latter as well. Intriguingly, this *fictional* contention that space and time are integrally connected preceded—by more than a decade—the first scientific assertions of the same point as suggested by Einstein's special theory of relativity.

Special relativity in and of itself permits only certain types of time travel, namely those directed toward the future, not the past. By traveling closer and closer to the speed of light, space voyagers' personal clocks would slow down relative to Earth time, allowing for indefinitely long voyages into the future. They couldn't, however, reverse course and return backward in time to the present.

Even if they traveled close to light speed in the opposite spatial direction, their voyage would still bring them farther and farther into the future.

A hypothetical particle, called a tachyon, is theorized to move forever at faster-than-light speeds and hence always travel backward in time. The logic behind this idea is that because moving closer and closer to light speed slows down time more and more, moving at light speed would stop time, and moving faster than that would make it run backward. However, in the more than four decades since this particle was proposed by the Columbia University physicist Gerald Feinberg, no such anomaly has been found. Moreover, an ordinary particle couldn't become a tachyon because it would take an infinite quantity of energy for it to reach light speed and another infinite amount to go beyond it, which obviously is impossible.

Does that mean that backward excursions through time would be out of the question? Not necessarily. General relativity offers far more flexibility than special relativity in that it allows space-time to curve in an endless variety of ways, depending on the precise configurations of matter and energy in a region. If space-time is twisted into just the right shape, it could enable what theorists call closed timelike curves (CTCs). Hypothetically, anyone who discovered one of these could travel, like Homer, back through the eons. Sweet.

The first known theoretical example of a CTC is a model of the universe proposed by the Austrian-born mathematician Kurt Gödel (pronounced "girdle") in 1949. The strange thing about Gödel's model is that it spins around a central axis like a carousel, unlike what astronomers believe to be actually the case. The astronomical consensus is that the universe is expanding, not rotating. No noticeable spin has ever been detected. One would think this would have proved a major hurdle (pronounced "Hödel") for the Austrian thinker, but he remained steadfast in his belief.

If space is like a whirling carousel, then time is like the mechanical horses rising on the poles. (In this analogy, we imagine the horses only rising, not falling, because normally we move only forward in time.) Picture the past to be the lowered positions of the horses and the future to be the raised positions. Because the horses are all oriented straight up and parallel, the "pasts" of each pole line up with the "pasts" of those next to it, and the "futures" of each pole line up with the "futures" of those next to it.

Now suppose that the poles are loosely attached to their bases. As the merry-go-round spins around, each pole might tilt and touch the one closest to it. The poles would no longer be parallel, but rather would be interconnected. By analogy, we note that the rotation of Gödel's carousel universe would enable contact between the future of each region and the past of its neighbor. This would allow continuous loops through time—in other words, CTCs. Hence, by traveling in any closed circle around the central axis of the universe, an explorer could journey backward in time. Theoretically, anyone with access to a powerful enough spaceship could attempt to change history.

Imagine, for example, that Moe wanted to go back in time, murder his grandfather, and eliminate all traces of his own miserable existence. He might figure that if he were never conceived he'd be off the hook for doing a lot of painful things, like being extricated from a womb, growing up, getting jilted, growing up some more, getting jilted again, cleaning up after Barney, getting jilted while cleaning up after Barney—the list goes on and on. It would be better, Moe might conclude, just not to bother with this world. Therefore backward time travel would be a way of putting that plan into action.

Suppose Moe started to plan out such a time-traveling, self-eradicating mission. He'd face a number of formidable obstacles, such as the enormity of the universe and the likelihood that it does not truly rotate (at least not enough to produce CTCs). Luckily for him, though, universal rotation is not the only potential source of CTCs. Other ideas include an infinite, spinning cylinder proposed by the Tulane physicist Frank Tipler in 1974, and a "traversable wormhole" system proposed by the Caltech physicists Michael Morris, Kip Thorne, and Ulvi Yeltsever in 1988.

The Caltech proposal has a curious history. Its time-travel method emerged out of a space-travel scheme imagining ways of connecting remote parts of the universe. Thorne originally concocted traversable wormholes as a way of fulfilling a request by his friend Carl Sagan, who required an imaginative, but scientifically feasible, way of having a character in his book *Contact* take a rapid interstellar voyage. Like mountain tunnels providing shortcuts between otherwise widely separated communities, wormholes are hypothetical tunnels through the spatial fabric linking up otherwise remote regions of the cosmos. In standard terminology, a wormhole has two "mouths" (entranceways), one on each end, connected by a long "throat" (the tunnel itself). The throat is carved out according to general relativistic principle by arranging just the right configuration of material, including a hypothetical substance with negative mass and repulsive gravitational properties, called exotic matter. Space travelers would enter the wormhole through a mouth located in one part of space, journey through the throat, and emerge in another mouth situated in a remote region of the cosmos.

As it turned out, not only could wormholes theoretically be used for space jaunts, under certain circumstances they could also be used for time travel. After the Caltech researchers sent off their scheme to Sagan, they noticed that a traversable wormhole could be transformed into a time machine by transporting one of its mouths at a near-light-speed velocity compared to the other. Following special relativistic time dilation would slow down the clock of the high-speed mouth compared to the low-speed mouth. While years passed for the former, only months might pass for the latter. Then if space travelers entered the low-speed mouth where many years had passed, journeyed through the throat, and popped out of the high-speed mouth, where only a few months had passed, they would be transported backward in time.

Let's see how this would work by picturing a scenario involving Kodos and Moe, and their grandparents. Imagine that Kodos's grandpod has constructed a wormhole time machine back in what we know as the 1940s, positioning one of its mouths in orbit reasonably close to Earth, and blasting the other mouth off into space on a roundtrip voyage at close to light speed. Consequently, while the first mouth has aged six decades, the second mouth has aged much less—a few months, say.

Now in the 2000s Moe has a strange vision of this wormhole and an unholy compulsion to commit a heinous act against his own flesh and blood. From leftover pull tabs, potato chip cans, discarded fissile material, and so forth, he builds his own spaceship and blasts off into space. As if in a dream, Kodos appears to him and helps guide him toward the wormhole's orbiting mouth—so carefully aligned by Kodos's grandpod, who was an orthodontist as well as an octopod. Moe's spaceship passes through the mouth, down the throat, and out the other mouth. He returns to Earth, but because of the second mouth's delayed time frame it is now just the 1940s. Moe, eyeing his grandfather, who is just about to head toward church and get married, doesn't wait long before ensuring the young groom is toast.

What Moe doesn't realize, though, is that (as we've pointed out) toast has a way of popping up again and again. By killing his own grandfather, Moe's parental line has been plugged and he should no longer exist. If he simply vanishes, however, who built the spaceship, went back in time, and performed the foul deed? Nobody. In that case, Moe's grandfather must have survived, gotten married, and had a child who engendered Moe. Thus Moe *does* still exist. In short, Moe is simultaneously alive and extinct—a fate arguably worse than being jilted by others, and more like constantly jilting himself. The bizarre scenario in which someone murders his or her own grandfather and thereby continues to pop in and out of existence is one of the most famous conundrums related to backward time travel and is called, appropriately enough, the grandfather paradox.

Science-fiction writers and others have pondered countless other contradictory situations that would be facilitated by traveling backward in time. Traveling forward in time wouldn't carry the same philosophical baggage because future history is, by definition, yet to be written and therefore can be changed in any way without creating a contradiction. The past, however, is a parchment scribed in indelible ink; like Bart's permanent record, it would not seem to be easily expunged. Thus it's odd to think of going back to those same pages and altering or obliterating what has already happened.

Because of such tricky conundrums, many scientists have argued that backward time travel is impossible. For example, *Simpsons* guest star Stephen Hawking, whose day job is Lucasian Professor of Applied Mathematics at Cambridge University, has argued for a "chronology protection conjecture," a theorem of physics that would preclude CTCs. The idea is that whenever someone tried to use general relativity to create a loop in time, natural forces would well up and destroy it, like rising tides smoothing over sand castles.

Thorne and his coworkers, along with the Russian theorist Igor Novikov, have taken a different tack. They've made the case that backward time travel is reasonable, as long as it is self-consistent. In other words, if someone journeys to the past and doesn't change history, but rather is a part of history, that's okay. The result, they contend, would be a coherent chronology of events, rather than one with messy, paradoxical twists.

For example, let's consider a time-traveling variation of the episode "Lisa the Iconoclast." In the episode, Lisa, while doing research at the Springfield Historical Society, discovers a confession note proving that Jebediah Springfield, the town's founder, was really an impostor. She informs the society's curator, Hollis Hurlbut, an ardent defender of Jebediah's legendary patriotic deeds, who attempts to cover up the truth. In the end, Lisa realizes that it's better for the town's spirit if no one knows what actually happened.

Now imagine, in this plot variation, that Hurlbut somehow comes upon a time machine and tries to determine once and for all the circumstances of Springfield's founding. He packs up some artifacts from the time, including the confession letter, which he puts in his pocket, sets the controls for the period when Jebediah Springfield was supposed to establish the town, and journeys backward in time to that era. Upon arriving at the site, however, Hurlbut finds no trace of Jebediah, nor does he meet an obvious impostor. Frustrated that the legendary historical events don't seem to be taking place, he takes it upon himself to carry them out. With his superb memory of history, he makes sure that everything that's supposed to happen really does, including a famous event where a wild buffalo is tamed. Later, he asks a local mason to carve Jebediah's name on a gravestone, digs up and dresses a body from the pauper's cemetery in a frontier outfit, and sets up a false grave. All this would be an undetectable replica except that just before Hurlbut returns to present-day Springfield the confession note drops out of his pocket and is left behind.

Upon arriving back in the present, Hurlbut finds to his relief that nothing truly has changed. Lisa still discovers the note and realizes that Jebediah was an impostor, but concludes wrongly about who he really was. The other townspeople still believe in the traditional legend. Thus, Hurlbut's time-traveling excursion has corresponded perfectly well to the historical record, offering an unambiguous, unified account of how Springfield was founded. Fitting in with the ideas of Thorne, Novikov, and others, the closed temporal loop he has enacted is absolutely self-consistent and free of paradoxes.

Nevertheless, the perplexing nature of the confession note raises a significant question. If an article discovered in the present is brought back to the past, left there, and eventually rediscovered, who originally created it? Apparently nobody. Therefore, its existence is an effect without a cause. Strangely enough, if backward time travel were possible, *anything* could be manufactured out of thin air.

For instance, suppose Smithers wanted to buy Burns a flawless, sparkling, five-pound diamond for his birthday, set on a ruby-encrusted, gold-plated emerald dish. All he would have to do is make the decision to get it, travel forward in time to when he's already given it to Burns, remove it from Burns's collection, and bring it back to the present. Then he could wrap it up and give it to

Burns. Enthralled by the gift, Burns would no doubt place it back in his collection, unaware that in some sense it had already been there. It would remain there until Smithers, in the future, removed it again and brought it backward once more in time. Clearly, no one has ever made the precious piece, yet it exists nevertheless. Self-consistency is no guarantee of reasonability.

Yet another alternative to forbidding backward time travel altogether, or insisting on rigid self-consistency, is postulating the existence of parallel universes. Suppose that whenever backward time travel changes the future course of events, reality itself bifurcates, engendering a whole new parallel strand. For example, during Homer's toaster-popping excursions, each voyage to the past would establish a completely independent timeline, with its own version of the Simpsons and Springfield. In some of these universes Flanders would ascend to dictatorship, in others he would just be a humble "neighborino," and in yet others he wouldn't even exist. Some realities would include plentiful donuts in handy cardboard boxes, in others donuts would drop from the sky, and in yet others donuts would be so rare and coveted that the human race would be perpetually fighting and biting to get them, sadly reduced to an Itchy-and-Scratchy-esque existence.

The concept of parallel universes has a basis in certain speculative physical theories, including the "many worlds" interpretation of quantum mechanics. This alternative to the Copenhagen (standard) quantum approach was proposed in 1957 by Hugh Everett, then a Princeton graduate student, and popularized by the physicist Bryce DeWitt. It mandates that every time a measurement that has more than one possible result is taken on the atomic level, physical reality splits up into a number of equally valid portions, one for each outcome.

The most famous application of the many worlds interpretation regards a conundrum known as Schrödinger's cat paradox. It imagines a scenario in which a cat is placed in a covered box and wired to an electron spin detector. Recall that spin is a quantum property for which an electron can have two possible values, called "up" and "down." If the spin detector reads "up" the cat survives, but if it reads "down" the cat joins Snowball I in kitty heaven.

According to the traditional Copenhagen interpretation, the cat remains in a mixed quantum state until an observer either reads the results of the detector or lifts the cover of the box. Only then is its state said to "collapse" into one of the two possibilities. In other words, roughly 50 percent of the time, curiosity kills the cat. The many worlds interpretation circumvents this issue by stating that the universe bifurcates into two branches. In one, the electron's spin is up and the cat alive; in the other, the spin is down and the cat deceased.

Could quantum physics ever produce a parallel version of Earth identical in most every way, except with no donuts? Because of its probabilistic nature, quantum mechanics allows for an almost unlimited variety of chance occurrences, including the unlikely possibility that the sugar molecules in every single donut spontaneously degrade into inedible substances—for example, methyl formate, which has the same chemical formula as the simple sugar glycolaldehyde but is used for insecticide. With such a "sweetener," bug juice would finally live up to its name, but donuts laden with hazardous chemicals would put off even Homer.

Time travel would be a risky game. Disrupting history would be bad enough, but imagine being trapped in an alternative reality with not even partially hydrogenated, calorie-packed crullers for comfort. Though we might be curious about the past and the future, most of us wouldn't want to take that chance. But what if we could view other times—events from long ago or many years from now—without having to step foot in those eras? How would we handle, for instance, seeing how our own lives will turn out many years from now? If the Simpson family is concerned, it might not be a pretty picture. Or as a certain bully might delicately put this, "I can smell your future. Ha-ha!"

16

Frinking about the Future

One of the greatest frustrations of life is the inherent unpredictably of the future. Natural forces are notoriously capricious, as the horrors of disasters such as earthquakes and tornadoes attest. Moreover, even if we could predict every aspect of nature, we'd have great difficulty anticipating the nuances of human decision making. Our planet is packed with billions of free-thinking individuals, each of whom has the power to change his or her mind at any moment. For that reason the lives of individuals and the histories of societies often veer in unexpected ways. A couple might spend all of their money to buy their dream house only to learn that their state has just voted to level that tract of land and build a highway. A woman might meet her ideal partner only to discover that he's just received his deportation notice. Or, in Patty's case, that his gender is not what she expected.

With *The Simpsons*, frantic plot twists and turns are par for the course. Watching the beginning of any episode, all bets are off as to where it will lead. For example, the episode "The Regina Monologues," opens with Bart finding a $1,000 bill (lost by Burns) and starting a museum with the bill as its basis. In a standard sitcom, running the museum would be grist for the chuckle-mill, and the show would end by resolving how Bart's wild scheme worked out. Not so with this series. The episode continues with the family going to England (with the money earned from the museum), meeting Prime Minister Tony Blair, author J. K. Rowling, and actor Ian McKellen, and bashing into the queen's carriage. Homer gets locked up in—and escapes from—the Tower of London. Grandpa reacquaints himself with a former lover and learns that he's the father of an illegitimate daughter who is the spitting image of Homer. Without seeing the coming attractions, who could have guessed all that from the episode's opening?

Despite this unpredictability, several episodes of the show concern themselves with foretelling the future. As in Matt Groening's other series, *Futurama*, the appearance of this theme seems to reflect his fascination with science fiction. Moreover, considering that for the ordinary episodes the characters never age at all (due to their being animated, but also a wise decision not to make them older artificially), the futuristic segments have offered the writers more freedom to flesh out the characters' lives. After all, having a two-decade-long series with hardly any changes to the principal characters is virtually unprecedented.

The three episodes centered mainly on glimpses of the future are "Lisa's Wedding," from season 6, "Bart to the Future," from season 11, and "Future-Drama," from season 16. The initial airdates of these episodes were spaced in five-year intervals, leading to my own prognostication that, assuming the show's still on, season 21 will be next. The pacing, however, is virtually the only thing foreseeable about the episodes. Consistent with the series' frenetic spirit and unpredictable nature, the episodes offer contradictory portraits about what happens to each of the principal players. Part of this stems from the muddling of timelines due to the shifting of which year is considered the present (for each year of real time, the "present" of the series moves one year later). For instance, "Lisa's Wedding" portrays the year 2010, when Lisa is supposed to be in college, and "Future-Drama" imagines life in 2013, when Lisa is said to be graduating two years early from high school! If the series keeps trucking along, however, by the time of the twenty-first season in 2010, Lisa will still be in second grade. (Unless she moves to Cletus the yokel's community, I doubt she'll be anywhere close to betrothal by then.) In the face of these conflicting parameters, the series' writers seem to suggest that because inherent limitations render all predictions moot anyway, the scenarios they present should be taken with a grain of salt.

In the three prophetic episodes, Lisa and Bart learn about their futures in different ways. In "Lisa's Wedding," Lisa encounters a gypsy fortune-teller, who seems to specialize in forecasting bad relationships. The method she uses to predict a catastrophically aborted wedding for Lisa is cartomancy, or card reading. In "Bart to the Future," Bart meets a Native American casino manager, who summons images of Bart's life when he is forty years old by using the method of pyromancy, or divination through fire. Flickering flames foretell Lisa becoming president, and Bart a major thorn in her side, à la Billy Carter. "Future-Drama" involves a machine invented by Professor Frink that he asserts is based on astrology, or interpreting the movements of the stars, planets, and other astral objects. Other methods of divination (to be reserved, perhaps, for future episodes) include phrenology, or reading bumps on the scalp; chiromancy, or palm reading (somewhat more challenging for four-fingered hands); cleromancy, rolling dice or casting lots; and oneiromancy, or the interpretation of dreams (perhaps the favorite of Springfield psychiatrist Dr. Marvin Monroe and his ilk). None of these methods has a scientific basis, despite Frink's claims about his astrology machine.

Frink's device works a bit like a DVD player or a TiVo with an onscreen menu brimming with choices. By highlighting and selecting an episode, one can view glimpses of life in the future for any of the characters. For instance, clicking on "Vice President Cletus" brings on a vision of Cletus the yokel asking his girlfriend, Brandine, to pack up his britches for an official trip to Brunei. For the bulk of the episode, Frink shows Bart and Lisa a more poignant subject: their lives around the time of their high school prom and graduation.

As the machine foretells, Bart and Lisa are each being pursued by narcissistic dating partners. A status-conscious girl named Jenda is pressuring Bart for an intimate relationship. He seems interested but afraid of losing his independence. At the same time, a pumped-up, muscle-obsessed Milhouse is manipulating Lisa, who has turned to him after he's saved her from a fire. Even after finding out that he actually started the fire, she remains with him apparently out of desperation. When a scholarship to Yale promised to her by Burns is retracted and offered to Bart instead, Milhouse exploits her feelings of dejection to bring her even closer. Meanwhile, after Homer squanders the family savings on an underwater house, Marge decides to separate and briefly date Krusty. This infuriates Homer to no end. The Snake Jailbird of time seems to have robbed the family of any chance for true love and happiness.

Eventually, Bart realizes that Jenda is wrong for him and that Milhouse is wrong for Lisa. The revelation comes after he comes upon Frink's house, now abandoned, and wants to see what's inside. Jenda leaves Bart in frustration when he wants to check it out instead of checking her out. Once inside Frink's laboratory, Bart finds the astrology machine and sets it even farther into the future. He sees a depressing image of Milhouse and Lisa's married life—with a distraught Milhouse informing Lisa that he has just sold all his bone marrow in a desperate attempt to pay their electric bill. Dismayed by how downtrodden his sister will become, Bart whisks her away from Milhouse and gives back the scholarship she rightly deserves. Marge and Homer reconcile, and all appears well on the Evergreen Terrace of the future.

Typical of the futuristic episodes, "Future-Drama" features a plethora of bizarre genetic mix-ups, weird robots, and malfunctioning technology. Moe now has an identical clone to help him out in the bar. In the cloning process, a spider with some of Moe's genetic material was inadvertently produced—a nod to the "Fly vs. Fly" episode, no doubt. This cloning reference is based on the popular conception that clones would be formed as full-fledged adults, rather than the reality that if the process were ever perfected, the clones would need to grow over time from single cells just like ordinary embryos. Thus Moe, if he had a clone, would more likely be changing its diapers than allowing it to tend bar.

Other forms of futuristic technology seem similarly screwy. Chief Wiggum is now a robot with a chicken rotisserie for a stomach. Homer's new underwater house requires three hours of decompression just to exit. He drives the first levitating car, which appears to be a lemon. Along with Bart, he traverses a "quantum tunnel" through a mountain.

For an elementary particle, quantum tunneling takes place in situations when its wave function extends through a barrier that the particle would classically be unable to cross. In that case, while classical physics would predict that the particle had zero chance of being on the other side of the barrier, quantum physics would rate that chance as small but finite. According to our current understanding, the quantum tunneling effect almost always applies to objects on the atomic or subatomic scales, rather than those the size of cars, but never mind.

When the car exits the mountain, it has somehow picked up Bender the robot from *Futurama*. (This is a rare case of crossover between the two series.) Bender's inexplicable appearance provides yet another example of the limits of future technology. When he tries to pal around with Homer and Bart, they seem completely uninterested and quickly toss him out of the car.

For a machine based on astrological forecasts, Frink's visions of the future seem incredibly detailed. Could real science do the same trick? We've seen how traversable wormholes, hypothetical shortcuts through space—for macroscopic objects, not just tiny particles—could offer tunnels into the past. An advanced future civilization, if it wanted to communicate with the past, could transmit streams of information through a wormhole in the hope that others could collect and interpret these messages. Ordinarily, such efforts would be risky, given that information about the future could change the course of history and potentially alter or even obliterate the civilization that sent the messages. However, if the future civilization were faced with imminent disaster, such as a planetary plague, an alien conquest, or a devastating nuclear war, its only hope might be a warning message transmitted backward to an age when the catastrophe might still be averted.

Gregory Benford's acclaimed novel *Timescape*, based on some of his own speculative ideas in theoretical physics, delves into such a planet-rescuing situation. A scientist from a future set in 1998 (the novel was published in 1980) develops a means of communicating backward in time in an attempt to warn those living in the early 1960s about impending ecological disaster due to certain manufactured chemicals that devastate the food chain. The mechanism used to send the message involves tachyons, which we recall are hypothetical particles that exceed the speed of light. Broadcast in modulated signals, like fluctuating radio waves, they are used to relay critical information into the past by interfering with nuclear processes in measurable ways. The scientist aims the signals in the direction of where Earth was in 1963, affecting the results of a nuclear experiment being performed at that time. When researchers from 1963 manage to decipher these results, they publish a key paper that prevents production of the damaging chemicals, thereby avoiding catastrophe.

Although tachyons have never been detected, and ordinary particles cannot be accelerated faster than light, nothing in the laws of physics strictly prohibits them. Thus, although Homer might preclude Lisa from violating the laws of thermodynamics in their household, he'd be loath to tamper with her sending tachyon signals warning of ecological catastrophe—a core meltdown at the nuclear plant, for example. She'd make a good case, in her soft but cogent voice, that relativity permits particles already traveling faster than light. Light speed acts as a barrier between slower- and faster-moving particles, the former known as "tardyons." Only crossing the barrier is forbidden.

Given these theoretical methods for relaying particles backward through time, it's curious why Frink bases his machine on astrology, rather than a true science. In making that choice he hearkens back to ancient times, when the line between astronomy and astrology was blurred. Early astronomers possessed more prognosticative skills than any other discipline of their day, with their vast storehouse of knowledge about the movements of stars and constellations (pictures assigned by mythology to various stellar arrangements), enabling them to chart calendars and predict celestial events such as eclipses. Presuming that their mastery of the celestial realm extended to earthly events as well, kings and other powerful figures turned to these "wise men" when important decisions needed to be made. Critical battles would be waged and important pronouncements rendered only when these astral advisers judged that the stars were correctly aligned.

Over the centuries, astrology has remained a lucrative and popular pursuit. Even Johannes Kepler, the seventeenth-century German pioneer of the scientific method, sold astrological forecasts to earn extra income. He shared the common misconception that the stars exert a steady pull on the course of human lives. In investigating celestial motion, he suggested that this knowledge might enhance our ability to prognosticate future events on Earth. Fortunately, he was able to put his preconceptions aside and focus on the story told by the data themselves. This led him to deduce the fundamental laws of planetary dynamics, a critical development that paved the way for Newtonian mechanics.

Thanks to Kepler and his Italian contemporary Galileo Galilei—who invented the astronomical telescope in 1609—astronomy has taken root as a modern science. Even if the stars don't hold the key to our personal fate, they provide essential clues about the origin and the destiny of the cosmos. And with the splendor of the skies open to our gaze every night, anyone from famous astronomers to curious eight-year-old girls has an opportunity to explore profound cosmic mysteries—assuming, that is, that light pollution doesn't get in the way.

PART FOUR

Springfield, the Universe, and Beyond

Has science ever kissed a woman, or won the Super Bowl, or put a man on the moon?

—Homer Simpson, Treehouse of Horror XV

There's so much I don't know about astrophysics.

—Homer Simpson, Treehouse of Horror VI

17

Lisa's Scoping Skills

Lisa Simpson's considerable talents are underappreciated by her family and schoolmates, save perhaps the longing, passionate gaze of Milhouse. Ah, unrequited love! A jazz musician, champion speller, amateur scientist, and custodian of the environment, Lisa is truly a renaissance pupil, a veritable Leonardo of the lunchbox generation.

What gall, then, has Eric Idle to appear on the show—in the guise of the character Declan Desmond—and accuse Lisa of being a "buffet-style intellectual" and a dilettante? While making a documentary about the kids in Springfield called "American Boneheads," he turns Lisa's assets into liabilities by mockingly calling her a "Jill of all trades" and snidely inquiring, "What's the ambition du jour?" He, the consummate comedian, comedy writer, actor, songwriter, member of both Monty Python and the Rutles, playwright of the Broadway musical *Spamalot*, among other endeavors, has some nerve pointing a finger at a sweet little girl for her lack of focus.

The episode in which Idle appeared, "'Scuse Me while I Miss the Sky," is about astronomy, of which he has ample experience in the musical vein. One of the classic bits from his late-stage Python days was a ditty he wrote and performed for the film *The Meaning of Life* called "The Galaxy Song." Sung to a frumpy housewife portrayed by Terry Jones, the song describes the meaninglessness of earthly existence in the unimaginable vastness of the cosmos. A compendium of astronomical knowledge, it points out Earth's minute place in our galaxy, the Milky Way, with its hundred billion stars, that, in turn, comprises but a minuscule fraction of our constantly expanding universe. Hence, compared to the endless sands of eternity, we are in essence but a mere speck of dust. Heavy stuff for a movie number.

If you've ever gazed at thousands of diamond lights pressed into the black velvet canopy of the sky, you may have experienced such a sense of minuteness. Standing in a pitch-dark field and looking up at the endless array of stars, you'd undoubtedly be awestruck by Earth's humble position among the enormity of everything. If you can't find a place in your region that's dark enough, wait until you are on vacation in a less illuminated locale—a campground, for instance. The stars will greet you like forgotten friends from a lost era. Or, if you still can't find a sufficiently dark place to stargaze, you might find your transcendental experience, like Lisa does, in the astronomy section of a natural history museum.

In the episode, following Declan Desmond's scathing critique, Lisa runs into the Springfield Museum of Natural History and embarks on a desperate search for her own identity. After spending time in several other scientific exhibits, including dinosaur and geology exhibits that are not particularly exciting, she comes across a spectacular planetarium show about our place in the universe.

This inspires her to want to be an astronomer. No more sampling from intellectual buffets; she wants to take on the main course.

Persuading Homer to buy her a telescope, Lisa sets out to explore the starry sky. Like Galileo did centuries before her, she hopes to peer at the planets and examine their curious features: Saturn with its prominent rings, Jupiter with its giant red spot and flotilla of satellites, the moon with its mountains and craters. Galileo, however, did not have to contend with the omnipresent glare of fast-food restaurants, shopping malls, twenty-four-hour convenience stores, traffic-clogged freeways, and so forth. When he was gazing through his instrument, the nighttime sky was absolutely dark, save perhaps for the soft glow of moonlight.

In contrast, Lisa's astronomical ventures must compete with a barrage of local light sources. A stadium's brilliant illumination overwhelms her efforts to view Venus, and the glare from the Starlight Motel kills any attempt to see Jupiter. Lisa flees to a hill but still can't escape from the "sickly orange barf glow" hovering above Springfield. Nearby is an astronomical observatory, run by Professor Frink, who confirms that light pollution is one of the biggest problems facing astronomers—even harder, he explains, than "getting a date."

Enraged by this issue, Lisa circulates a petition and convinces Mayor Quimby to dim Springfield's evening lighting. Townspeople are amazed by the spectacle of starry patterns. Lisa looks forward to an upcoming meteor shower that she hopes to view in its full glory.

The darkened skies, however, prove a boon for criminals—particularly vandals who like to saw off hood ornaments from cars. Even Bart and Milhouse, trying to look cool, join in on the hood ornament craze and try to snag one of their own. The ensuing public outcry forces the Mayor to "flip flop" and crank up the town's lighting even brighter than before—to the level of "Permanoon"—foiling Lisa's astronomical ventures as well as Bart and Milhouse's shenanigans.

Now, instead of vandalism, insomnia has become the key issue. With a flood of electric illumination seeping into every nook and cranny of Springfield, no one can get any sleep. Homer is absolutely catatonic, which proves highly convenient for Lisa and Bart, who hatch a scheme to darken the sky. Homer, in his hypnotized state, is led by Lisa and Bart to the nuclear plant and is compelled to deactivate the security system. Lisa and Bart then set the plant's output switch to overload and burn out all the lighting in Springfield. Suddenly all the power is out and the glow no more.

Against the ebony backdrop of the sky, Lisa, Frink, and the others are able to admire the wonders of a spectacular meteor shower. Frink inspects a fallen meteorite and finds evidence of the carbon-based molecules needed for life, until this proof is whisked away by a tiny hitchhiking alien. No matter—at least for the time being, all is right with the night.

Grappling with light pollution is one of the challenges of contemporary astronomy. A century ago, research observatories could be placed almost anywhere, even in the suburbs of major cities, and still take advantage of skies dark enough to deliver critical information about the cosmos. One of the greatest discoveries of all time—the expansion of the universe—was made at Mt. Wilson Observatory, less than twenty miles from downtown Los Angeles. There, with the hundred-inch-diameter Hooker telescope, Edwin Hubble determined the distances to numerous galaxies, establishing conclusively that they lie well beyond the Milky Way and, moreover, that they are moving farther and farther away from us (and each other). These discoveries were made at a time (the 1920s) when Los Angeles was already a major city and principal center for movie production, yet the skies on Mt. Wilson were dark enough for Hubble to collect light from variable stars in galaxies millions of light-years away (one light-year, the distance light travels per year, is about six trillion miles). By recording the observed brightness of these variable stars, called Cepheids, compared to how much energy they were actually producing (a known quantity for that kind of star), Hubble was able to figure out how far away they were and hence the distances to their host galaxies. Combining these data with information about each galaxy's outward speed, he demonstrated that the galaxies are fleeing and that space is getting bigger and bigger. All these critical observations took place merely a film canister's toss from star-speckled Hollywood.

Today, although there are still many observatories near cities, their scientific usefulness has been greatly reduced by urban haze. Instead, researchers clamor to book time on large telescopes in the mountains of Chile, atop the highest peak in the Hawaiian Islands, and other far-off regions. To completely eliminate light pollution and atmospheric distortion, numerous probes have been propelled into the vacuum of space, including the Hubble Space Telescope, named after the great astronomer and launched in 1990.

Free from the haze and blaze of Earth's atmosphere, the Hubble Telescope has been extraordinarily successful in imaging the farthest imaginable reaches of the cosmos, collecting the light of numerous galaxies billions of light-years away. When in 1995 a tiny, seemingly barren patch of sky observed by the Hubble in its Deep Field survey revealed thousands of galaxies just in that region, astronomers realized that space contains more than fifty billion galaxies. That's considerably more than Martin Prince could write about in his book reports, even if he drank a hundred all-syrup Squishees a day for his whole elementary school career. Not that he would go that route, mind you.

Hubble has since pushed the boundaries of astronomical knowledge ever outward, extending our understanding of the cosmic past closer and closer to the dawn of time. Because light takes time to reach here, the greater the distance a telescope is imaging, the farther back in time it is probing. So, for example, when we view a star that is sixty-five light-years away, its rays have taken sixty-five years to reach us, and therefore we are seeing what it looked like at a time on Earth when Abe Simpson might have been chasing army nurses. Although by today such a body could have lost most of its sizzle and settled into a state of quiescence, back then it could have been red-hot and smokin'. Peering back to the time of World War II or even back to the age of the dinosaurs, however, is trivial compared to Hubble's feats. Hubble has produced images of objects so far away that their light was produced during the first 5 percent of the universe's history—the first half of the first inning of the whole ballgame.

Aside from charting the depths of space and time, Hubble has revealed a veritable bling box of cosmic marvels: the jewellike patterns of planetary systems in formation, the distinctive afterglow of the incredibly energetic blasts known as gamma-ray bursts, the dusty residue of colliding galaxies, and so many other incredible images. It has even pointed to telling absences: places where black holes ought to be, gaps where invisible matter is thought to reside. No wonder astronomy captivates children like Lisa who are curious about the myriad wonders in the heavens.

Hubble, though the most famous space instrument, is not the only one. In recent years, it has served as part of an ensemble, working in harmony to cover the full range of the light spectrum. Like a string quartet with its lilting violins, midrange viola, and sonorous cello, NASA's great observatories of space span the highest frequencies (rates of oscillation) of light as well as the middling and low. The Compton Gamma Ray Observatory, launched in 1991 and operating for more than nine years, collected light of frequencies so high that it was well beyond visibility. This is analogous to dog whistles having such high pitch that they are beyond human audibility.

As quantum physics informs us, a light wave's frequency is deeply connected to its energy. Therefore gamma rays, the highest frequency light, are also the most energetic. Fortunately for living organisms, they are blocked by Earth's atmosphere and far easier to detect in space—a prime motivation for the Compton. The information collected by the Compton, in tandem with optical results from Hubble, has yielded vital clues about catastrophic events such as fantastically powerful stellar explosions.

Recording images of somewhat lower frequency, but still beyond visibility, the Chandra X-ray Observatory has filled a vital niche between the Compton and Hubble. X-ray signals are churned out by many energetic processes, such as black holes gobbling up nearby matter. As the material falls into their infinite gravitational wells, X-rays stream off into space and can be collected by detectors. Thus, although we can't see the black holes themselves, Chandra has recorded ample evidence of their voracious appetites. Chandra has also furnished proof of the existence of intergalactic dark matter.

Completing the intended quartet (which has, alas, become a trio due to Compton's demise in 2000), the Spitzer Space Telescope measures light in the infrared range, with frequencies too low to see. Infrared radiation is better known as ordinary heat, the sort given off by human bodies and warm donuts, and appetizing relationships between these. Picked up by special night-vision goggles, it offers kids like Bart a chance to watch their dads ingesting forbidden donuts in darkness. Although stars like the sun radiate mainly in the visible range, planets such as Earth produce only infrared light. They can reflect visible rays from a star, but emit just infrared.

Given Spitzer's capabilities, it's not surprising that its major triumphs include imaging planets that are impossible to see optically because they are so far away. In 2005, Spitzer provided the first direct pictures of planets in other stellar systems, proving beyond any measure of a doubt that the solar system is not unique. While those purportedly kidnapped by Kodos, Kang, and their ilk never doubted this premise, Spitzer's findings clued in the rest of us.

Although space telescopes have produced magnificent results so far, they cannot be the only answer to the problem of light pollution. Extremely expensive to build, launch, and maintain, each requires decades of planning, budgeting, and political wrangling. When systems fail, as they are bound to, they must either be abandoned or have their use curtailed, or spacecraft such as the shuttle must be sent to make repairs. The shuttle program is due to be phased out, presenting major problems for the long-term maintenance of observatories in space.

What, then, is to be done about keeping earthbound observatories as free as possible of the nighttime glow that characterizes modern urban and suburban life? The first step is bringing the problem to public recognition. Like the citizens of Springfield, many people associate bright street lamps with safety and anything less with danger. Many don't realize that glare, the result of brilliant lighting alternating with relative darkness, represents a hazard in and of itself. It can actually reduce visibility, rather than enhance it. For example, suppose a respectable elderly citizen—say, the wealthy owner of a nuclear power plant—is strolling down the main street of a town and encounters a group of ruffians. If he has just passed a bright street lamp or a car with high-power headlights, his eyes might not have time to adjust, leaving him temporarily blinded as he attempts to maintain control of his briefcase full of gold doubloons. Thus even a baron of energy would be wise to advocate responsible public lighting schemes.

The noted amateur astronomer John Bortle, writing in *Sky and Telescope*, has proposed a nine-step dark-sky scale to evaluate the suitability of an area for astronomical observation. The scale uses the visibility of the main band of the Milky Way galaxy as a gauge of nighttime darkness. Level 1, the absolute best viewing conditions, represents as close to total darkness as possible. No nearby earthly objects can be seen—not even the telescope perhaps—as the sky and everything around are jet black. Under those circumstances, the Milky Way appears as an unmistakable creamy cloud stretched out across the sky. Level 9, the worst possible conditions, corresponds to the "Permanoon" setting of Mayor Quimby's dial, or something close to it, in which virtually no stars can be seen with the naked eye and perhaps only the moon and a planet or two are discernable. The brilliantly lit Ginza district of Tokyo, the Times Square area of Manhattan, and other concentrated urban regions are all Level 9. In between are rural and suburban skies of various degrees of illumination. Where does Springfield fit in? It seems to depend on which segment of the population Quimby is trying to pander to that evening. If it's his girlfriend, you'd better believe it's one of the darker settings—for the public good, of course.

The International Dark-Sky Association is an organization dedicated like Lisa to reducing light pollution around the globe. It strongly advocates replacing fountainlike street lamps that spray light in all directions with glare-free units specially designed to illuminate only the places under them. That way, people can clearly see the streets and sidewalks, but are not blinded by excessive lighting. Moreover, by eliminating the wasted light shining upward, communities are able to reduce the "orange barf" glow above them and enjoy the stunning vista of the Milky Way instead.

Light, of various frequencies, rains down on Earth all the time, offering a wealth of information about the universe. Astronomers strive to collect and interpret as much of these data as possible, typically presuming that the light has taken a direct course from the star or galaxy that produced it to Earth. As Albert Einstein showed, however, this assumption is not always valid. The gravitational influence of intermediate objects can bend the path of light and create curious optical illusions—multiple images of the same astronomical body. This strange phenomenon, a key prediction of general relativity, is called gravitational lensing. It's well known to astronomers, yet people unfamiliar with its effects, such as the Simpsons, can mistake it for clinical insanity.

18

Diverting Rays

With the series *Everybody Loves Raymond* so popular and its star, Ray Romano, so affable, the *Simpsons* writers found Ray diverting enough to craft an episode around him. Unfortunately, Ray turned out to be so diverting that even light bent around him, so that for most of the segment nobody but Homer could see him. Diverting rays of light can offer splendid optical illusions, lucrative magician's stunts, means of testing astronomical theories, and all the social advantages of invisibility. In Homer's case, his family and friends just considered him utterly mad.

The episode's title, "Don't Fear the Roofer," is a play on the Blue Oyster Cult song "Don't Fear the Reaper," which inspired an entire generation of air-guitarists and enraged an entire generation of bible-thumpers. Roofers can be scary indeed, swooping down in the dead of night, forcing their way through loose shingles, and snatching away unlucky souls. Then they return the next day with a hefty bill.

Ray Romano plays a more amiable type of roofer named Ray Magini, who offers to help Homer to fix his leaky roof for free after it has been pummeled by a storm. They meet each other at a bar and are on the road to becoming fast friends as they share beer, nachos, and plans to fix Homer's roof. However, the plans go stale when Homer and Ray end up goofing off with a rooftop nail-gun fight rather than attending to the job.

When Marge discovers that the roof still isn't patched, she's furious. She implores Homer to do the job on his own. Along with Bart, he scurries down to a home-supply store called Builder's Barn, where he runs into Ray, who assures him that he'll soon return to the Simpsons' house and fix the roof. Later at home, Homer waits some more for Ray, until Marge begs him to stop and advises him of her opinion that Ray is just a figment of his imagination. Bart concurs, stating that in Builder's Barn it looked like Homer was just talking to himself. The clincher is that Ray's full name is an anagram for "imaginary."

Homer is then hauled off for electroshock treatment with Dr. Hibbert. Jolted again and again, Homer finally stops believing that Ray exists. Then Ray shows up, startling the family, and making Dr. Hibbert nervous about a potential lawsuit. Now that everybody sees Raymond, they wonder why they didn't before.

Besides Ray Romano, the other guest star of the episode is Stephen Hawking. In the second of his two appearances on the series, Hawking portrays himself as the new owner of a neighborhood Little Caesar's pizzeria and pretends that his computer-generated voice becomes stuck while saying "Pizza, Pizza!" Being an expert in astrophysics, Hawking furnishes a grand relativistic explanation for why Bart didn't see Ray at Builder's Barn. Hawking reports that he's "been tracking a tear in the fabric of space-time, which combined with airborne pieces of metal at Builder's Barn to create a miniature black hole . . . between Homer and Bart, causing a gravitational lens, which absorbed the light reflected from Ray the roofer."

Hawking's explanation receives a nod from Lisa, but is it truly feasible? Gravitational lensing is a well-understood physical phenomenon that stems from the curvature of space-time by massive objects. When matter warps the fabric of space-time around it, straight lines, such as the paths taken by light, bend into curves. When the eyes trace these paths backward through space, they see distorted images. Einstein first described this effect in 1936, when he posited that if two astronomical bodies—two galaxies, say—are situated such that one is in front of the other on a line of sight toward Earth, the light from the farther object would be bent around the nearer object in all different directions, forming a ring. Such "Einstein rings" have been observed by the Hubble and other telescopes as partial or complete bull's-eye patterns of light.

If the bodies aren't perfectly lined up, they can still produce a lensing effect, albeit not a ring, but rather multiple images of the hind object. This effect was first observed in 1979, when astronomers using the Kitt Peak National Observatory noticed a pair of twin quasars (distant ultra-powerful light sources of galactic mass) that appeared identical in every way except that one was a mirror image of the other. They soon deduced that it was the light from the same quasar, split by an intermediate galaxy into two images. Since then, numerous other examples have been identified.

Gravitational lensing occurs for smaller objects too, not just those of galactic size. However, the less massive the lensing object, the less noticeable the effect. For example, if a remote planet from our galaxy passes in front of a star from another galaxy, it could bend the light from the distant star by a minute amount, target a fraction more of the star's rays toward Earth, and thereby cause the star to appear slightly brighter during the planet's interval of passage. For a number of years, scientists have tried to detect unseen planets using this technique, called gravitational microlensing. Several promising candidates have already been located in this manner.

Black holes could cause microlensing too, but in space, not on Earth, and certainly not on the minuscule scale mentioned in the episode. A black hole formed by pieces of metal lumped together in a home supply store would be extraordinarily tiny and virtually impossible to produce, even with the help of a "tear in the fabric of space-time." Typically, black holes are forged during colossally energetic implosions of stellar cores, when the gigantic stars containing these cores suddenly blast apart in events such as gamma-ray bursts. When these cores implode, the atoms within them are absolutely pulverized. No form of store-bought material could create such extraordinary conditions.

Even if a "rip in space-time" could somehow cause such a calamity, it would undoubtedly affect the region around it. Why didn't anyone in the store notice an implosion of metal? Moreover, if one estimates the size of a black hole formed by such a collection of mass, it would be far too small to detect. Its bending of light would be completely unnoticeable and certainly would not affect Bart's ability to see Ray. Maybe Ray—or at least the metal floating around him—isn't that diverting after all.

But Ray is a crafty sort, so maybe he has crafted another means of invisibility, unbeknownst to the Simpsons (and even the writers of the show). To play a joke on Homer, perhaps he has been tinkering with experimental methods of rendering himself unseen. Like the Invisible Man from H. G. Wells's novella, he's trying his best to blend in.

Let's suppose that Ray has stumbled across an article by Professor John B. Pendry of Imperial College, London, a modern expert on the science of invisibility. By following Pendry's advice, Ray has developed a diversion scheme effective enough to block Bart's view. A bit of a stretch perhaps, but no more so than black holes made of commercial sheet metal.

Pendry's theories, published in *Science* and other well-known journals, involve using specially designed substances, called "metamaterials," to redirect electromagnetic fields (light rays) around objects and return them to their original paths. In other words, the light continues on as if the intervening objects weren't there. Pendry and his colleagues have calculated that such diversion is possible in principle and hope to complete an experimental design in the near future.

Now imagine that Ray Magini has already concocted such a metamaterial from various items found in the packed aisles of Builder's Barn. By stacking slabs of this material around himself on all sides—save the side facing Homer—he could play a trick on his newfound friend. Light from the rest of the store would slide around him like a curveball, returning to its original trajectory and continuing as if nothing interceded. Consequently nobody except Homer—whose line of sight is not masked by the metamaterial—could see him.

Creating the illusion of invisibility on a simpler level is a common magicians' trick. Magicians aren't concerned with whether you can actually see through an object; they just want to make you think you can see through it. In that case, carefully placed mirrors can usually do the trick. So if Krusty wanted to make Sideshow Mel disappear, he might ask him to duck into a box with a slanted mirrored front. The slanted mirror could reflect a pattern on the ceiling that is identical to a pattern behind the box, leading viewers to believe that Mel had vanished.

Ordinary mirrors and lenses bend light when it either reflects from (bounces off of) or refracts through (crosses) the boundary between two different materials—for example, the interface between air and glass. As light reaches such a boundary, it changes its speed and begins to take a different course through space. Hence, well-placed mirrors can bend light enough to make a box look like it is empty when it is really full.

Optical means of bending light are far more common and effective on Earth than any discernable diversion of light by gravity. Unlike standard lensing, gravitational lensing is noticeable only as a deep space phenomenon over extremely long (interstellar) distances. Indeed, in general, it is far easier to use electromagnetic properties than gravitation for just about any kind of bending or manipulation. For one thing, while gravity is always attracted, electricity and magnetism can be either attractive or repulsive. Moreover, electromagnetism is a far, far stronger force than gravity. You can see this by picking up an iron thumbtack with a small household magnet. The strength of the magnet, in lifting the thumbtack, overcomes the entirety of Earth's gravitational force trying to pull it down.

Considering their relative strengths, it's not surprising that while electromagnetic waves (that is, rays of light) are extremely easy to discern, gravitational waves have yet to be seen. They are produced by similar mechanisms: the former by oscillating charges and the latter by oscillating masses. Yet electromagnetic waves are much easier to produce and detect, as seen in conventional radio and television broadcasting.

When Kent Brockman broadcasts the news, a microphone picks up his voice. Sound causes a diaphragm within it to vibrate, jiggling a magnet, creating, in turn, a varying electrical signal. The operating principle is that whenever magnets move near a wire they generate an electrical signal. Similarly, Brockman's image is videotaped and transformed into another signal. These audio and video signals are combined and sent to a large broadcast antenna. The varying signal causes charges within the antenna to oscillate and produce changing electric and magnetic fields—in other words, an electromagnetic wave. Simply put, shake up charges and you make a wave. Shake them to the pattern of your voice and image, and the wave is shaped by your voice and image.

Now let's look at the receiving end. First, let's set aside cable and focus on broadcast TV. Imagine that Burns has bought up all the cable companies in Springfield and is charging $1,000 a month to get standard service. D'oh! We all know that not having TV makes Homer go crazy, so suppose he asks Ray to install a giant rooftop antenna. An antenna works because it contains charges that wriggle and jiggle in beat with incoming radio waves—low-frequency types of electromagnetic signals. When these charges oscillate they produce a varying electrical voltage that controls, in turn, the sound and the pixels (points of light) on a television set. Presuming the set is tuned to the right channel, with each channel representing a different frequency of radio waves, Kent Brockman will appear, or perhaps Itchy and Scratchy. The whole family can then bask in the warm glow of a sunny news item or a violent cartoon. Ah, the miracle of television.

For some physicists, however, electromagnetic waves are passé; they are *so* twentieth century. Trying to catch elusive gravitational waves is the new sport. Researchers involved with the LIGO (Laser Interferometer Gravitational-Wave Observatory) project, for example, have set their sights on bagging gravitational radiation streaming in from deepest space. Established by groups from Caltech and MIT, LIGO maintains detectors in the states of Louisiana and Washington, hoping to record the faint "broadcasts" of gravitational signals. These would be picked up not by oscillating charges, but rather by vibrating masses.

Unlike Springfield's nuclear plant, the LIGO project requires delicate and responsible monitoring. That's because the cascades of gravitational waves strong enough to be detected would be produced only in rare cataclysmic events such as a supernova explosion—when a giant star expels most of its material in a powerful outburst of energy—or a collision between two black holes. Even so, the test masses suspended within the LIGO detectors, designed to vibrate under the influence of gravitational waves, would jostle only a minute amount in response—less than one-trillionth of the size of a grain of sand. Amazingly, the LIGO instrumentation is sensitive enough to measure such disparities. However, viable candidate events that have no other explanation have yet to be found. For example, one candidate for a gravitational disturbance turned out to be an airplane flying overhead. Undaunted by the dearth of definitive findings, the LIGO researchers continue to search through reams of data, hoping someday to identify the unmistakable gravitational signature of a catastrophe in space.

If gravitational "broadcasts" are ever detected, they might prove more interesting than the typical television fare in Springfield.

Imagine tuning into live coverage of a supernova's blast; the explosive antics of Itchy and Scratchy would seem tame by comparison.

As we can see from the LIGO experiment, some phenomena in physics and astronomy are subtle, requiring extremely sensitive instruments to detect. Other occurrences are much easier to observe, even presenting themselves in household situations. Take, for example, the common household toilet. Could ordinary flushes reflect the influence of planetary spin or do you need a more delicate instrument? In a visit to the land down under, Bart aims to find out.

19

The Plunge Down Under

Australia is a land of many wonders, from the duck-billed platypus—an improbable egg-laying, pouched, beaverlike, webbed-footed mammal—to the conservative members of parliament known as Liberals. The weather there is also topsy-turvy. Christmas in some of the northern and central regions is typically more than 100 degrees Fahrenheit. When Santa bungee-jumps down from his sleigh, as he makes his run over the Northern Territory, he often dives into whatever pools or creeks he can find to cool off. It's not a pretty sight when all the gifts are soaking wet and covered with crocodile scales. Some folks swear that the water in Australia is different, too. I'm not talking about what comes out of the taps, which is called lager, but rather what swirls around in what they call the "dunny." That's what other parts of the globe refer to as the "john," the "loo," the "WC," or "Out of Order."

The episode "Bart vs. Australia" begins with Bart and Lisa pouring shampoo and toothpaste down the drain of their bathroom sink and watching it flow counterclockwise. Lisa claims that draining water always swirls counterclockwise in the northern hemisphere and clockwise in the southern hemisphere due to a physical property called the Coriolis effect. Thinking that preposterous, Bart flushes a toilet and sees the water twirl counterclockwise just like it does in the sink. Then, to check what happens in the Southern Hemisphere, he calls people in several different locales, including Antarctica, South America, and Australia. When he phones Australia, he reaches a town called Squatter's Crog, where a boy named Tobias picks up. Tobias checks his own sinks and toilets and those of his neighbor well down the road, and confidently reports that that they all drain clockwise. Based on these reports, Bart reluctantly concludes that Lisa was right. Sounds like the draining effect is fair dinkum (for real)—or is it?

A mishap brings the Simpsons down to Australia to see for themselves. Tobias has left his phone off the hook the whole time he is checking the appliances, resulting in a massive phone bill. Because Bart has called collect, Tobias's father, Bruno, is stuck with paying the tab. Furious about the phone bill, he reports Bart to a collection agency. After Bart fails to respond, the Australian government complains to the American government. To preserve the delicate relations between the two countries, Bart is faced with two alternatives: prison or a public apology in Australia. The choice is easy; the Simpsons hop a plane to the land down under.

Once in Australia, the family stops at the American embassy, where the toilets are specially rigged to flush "northern hemisphere style." Although upon flushing they first run clockwise, the water is forced into the counterclockwise direction to make Americans feel more at home. A growing flood of evidence seems to bolster Lisa's premise—a deluge augmented by each flush. But is this torrent truly palatable, given its source?

The answer is negative, as anyone who has studied the water from toilets (eau de toilette is the technical term, I think) might attest. Under ordinary circumstances, drains don't really run in reverse in the southern hemisphere. Yes, there is a hemispheric effect that can influence the flow of water, but for a small basin the size of a sink or a toilet it can be seen only under extremely precise laboratory conditions—with absolute stillness and total symmetry—not through normal hygienic practices. Then why, in the episode, did Lisa—who is supposed to be the smart one—claim otherwise?

Cartoons can help us understand and appreciate science, but they often exaggerate or distort the properties of nature to get a chuckle. Witness scenes from the Road Runner cartoons where the villainous Coyote hangs above a canyon for many seconds in the air while contemplating his doom; only after he and the viewers have had ample time to think about his fate does he finally plummet to the ground. As Roger Rabbit explained in his debut feature film, the only unbending law of cartoons is to get a laugh (unless you count O'Donnell's "Laws of Cartoon Motion," discussed in chapter 9). Nevertheless, in parodying physical principles, cartoon scriptwriters fundamentally need to understand them and often end up doing a lot of homework in preparation.

In "Bart vs. Australia," the show's writers deliberately set out to make Australia and the southern hemisphere seem as bizarre as possible. To tap into a vein of humor that derives from cultural differences, real or perceived, they purposely played up a number of popular misconceptions—for example, the "Crocodile Dundee" image of Aussies being unfamiliar with modern life, when in reality the majority live in urban or suburban areas. Therefore, highlighting an urban legend about toilets swirling in the opposite direction fit snugly into the writers' plans.

Some folk myths have a grain of truth; for instance, the real Dracula drew blood through his murderous actions but apparently didn't drink it. Pop Rocks crackle when mixed with soda but can't make you explode. A pale, gaunt bogeyman does haunt the dark streets of Springfield on Halloween night, but his trusty assistant makes sure that he doesn't harm any children.

In the case of the legends swirling around swirling, the popular misconception is rooted in actual large-scale effects due to Earth's rotation that are discernible in cyclones and hurricanes rather than in sinks and toilets. It's like assuming that because an elephant and a mouse are both four-legged creatures they share the same gait. What applies to the massive doesn't always appear in the tiny.

The Coriolis effect, in which fluids or otherwise unhindered objects (a free-moving pendulum, say) tend to deflect from their original positions over time in a clockwise or counterclockwise motion, stems from the Earth's surface being a noninertial reference frame compared to the "fixed" space around it. A noninertial reference frame means one in a state of acceleration: speeding up, slowing down, or rotating. Newtonian physics informs us that the laws of motion appear different from such a perspective than from an inertial (at rest or a constant velocity) point of view. The reason for this difference lies in Newton's famous principle of inertia: objects at rest tend to remain at rest, and objects in motion tend to keep going at the same speed in the same direction, unless an external force compels them to change their path.

The principle of inertia can be seen whenever Marge pushes Maggie in a frictionless stroller through an immense, perfectly flat roller rink. If Marge becomes distracted and lets go, Maggie and the stroller will keep going at the same rate along a perfect straight line indefinitely. Anyone standing in the roller rink would share the same inertial perspective and agree that Maggie's motion is uniform and linear. Even if Bart is moving at a constant velocity on his skateboard, he would concur.

However, suppose hapless Hans Moleman has decided to take up skating and finds himself in the peculiar situation where he starts to pirouette out of control. He'd then be in a noninertial reference frame. As he spins around and around, if he glances at Maggie's stroller he might think she is moving in a kind of spiral away from him, rather than in a straight line, largely because nothing looks like a straight line to someone spinning.

Now let's picture the opposite situation. Suppose the entire Springfield roller rink lies on a turntable, like a revolving restaurant, and slowly spins on its axis. Imagine that at the very center of this turntable, there's one place that doesn't turn—a fixed platform where Moleman is perched. Right in front on him, but on the rotating part of the rink, is Maggie's stroller. If Maggie's stroller is pushed and let go, Moleman would now be in an inertial reference frame. He would see the stroller moving in a perfectly straight line away from him. However, Marge, Bart, and the others on the rink would be spinning around in a noninertial frame. Instead of seeing Maggie move in a straight line, according to their rotating perspective she would seem to be swerving.

What causes Maggie to swerve? A stationary outsider, such as Moleman, would realize that she is actually following the normal course of inertia. In the absence of external forces, nature prefers straight-line motion. However, someone on the rotating platform itself, such as Marge and Bart, might not know this. From their perspective, an extra force, or set of forces, is pushing on Maggie's stroller and bending its path. Such forces are sometimes called "fictitious" because they can be explained away by inertia. They include what are called the centrifugal and Coriolis forces.

Many classical physics texts discount these noninertial forces because of their disappearing act once a different perspective is taken. As Einstein and others pointed out, however, who's to say which is the ideal perspective—a spinning or a stationary framework—because there is no absolute vantage point as a basis of comparison? Earth, for example, is rotating about its axis and revolving around the center of the solar system, which, in turn, is spinning around the center of the Milky Way. Einstein's quest to incorporate all possible frameworks into a single theory was one of his motivations for developing his general theory of relativity, mentioned earlier.

Of the noninertial forces, the centrifugal force is probably most familiar. It is the feeling that you want to move outward when you are on a spinning object. For example, when sitting on the school bus, each time Otto swerves, little Ralph might feel like he's about to go out the window, but fortunately the glass pane—which acts as a centripetal (center-seeking) force keeping him in—blocks him from flying outward. Those inside the bus might conclude that the centrifugal and centripetal forces balance. However, those outside the bus would just see one force, the centripetal, causing the kids to swerve back and forth with the bus's motion.

The Coriolis force is a bit subtler. It kicks in when an object moves closer to or farther away from the axis of a spinning body, such as the North and South Poles of the Earth, due to differences in speed between the two regions. Take, for example, a northerly ocean current starting in the tropical climes of the Caribbean Sea. It begins its journey not only traveling northward, but also moving eastward with the Earth at the same rate as the Earth's turning. As the current ascends toward the North Pole, it preserves the same eastward speed. However, points of more northern latitude don't need to cover as much ground in the same time period (twenty-four hours) as those near the equator and therefore move slower. Consequently, in more northerly regions the current's eastward speed outpaces Earth's, and the current turns increasingly in that direction. Eventually, it moves due east and traverses the Atlantic Ocean to Europe. This pushes the waters near Europe south, where they travel slower than the Earth's rotation and thereby move west. The end result is a clockwise movement of water, known as the Gulf Stream, serving as a conveyor belt for bringing a measure of Caribbean warmth to the northern European coast.

In the southern hemisphere the Coriolis effect works in the opposite direction, because the farther south a current moves, the faster eastward it is going relative to the Earth's rotation, causing it to veer counterclockwise. In other words, while north then east is clockwise, south then east is counterclockwise. This difference is noticeable not only in ocean currents, but also in major weather events such as hurricanes and tropical storms.

Unless you have an exceedingly large circular bathtub, however (the size of a lake, let's say), it's doubtful you would notice this effect during ordinary draining. The difference in the speed of Earth's rotation in two parts of a standard basin corresponds to their difference in latitude—which is to say, it's minuscule. Other factors in draining, such as any lingering currents in the water left over from when it was filled or asymmetries in the tub, play far more important a role. Hence, only if the tub were absolutely still and absolutely symmetrical could scientists under laboratory conditions measure the Coriolis effect's minute contribution to draining, and reportedly researchers have done just that.

It's a shame that when Bart and Lisa visited Australia, they didn't stop by the University of New South Wales in Sydney. Working there in the physics department is professor Joe Wolfe, who has written much about the Coriolis effect. He asserts that by filling up a basin, letting the water settle for a long time to eliminate any residual currents, and then carefully pulling out the plug, he could demonstrate (to curious children like Bart and Lisa, for example) how the water would run out with no rotation in either direction. Hence the Coriolis effect would not affect its draining.

For a household bathtub not so carefully prepared, Wolfe explains:

> [T]he direction in which it drains might depend on the location of the tap you used to fill it, because that can set up a circulation pattern during filling. If you have hot and cold taps on opposite sides, you might get different results for hot and cold water! Also, some basins might not be symmetric, so in some basins you might tend to get more than 50 percent clockwise, while others would be less than 50 percent. Nevertheless, these effects should cancel out. People who have done the experiment in the U.S. report, on average, 50% each way. That's not what you would expect. But people often confuse what they expect to happen with what really does happen.[1]

Wolfe is quick to point out that although typical bathtubs and sinks wouldn't offer the opportunity to observe the Coriolis effect, a device set up in the lobby of the University of New South Wales's physics building would do the trick. Called a Foucault pendulum, after its inventor, the French physicist Jean-Bernard-Léon Foucault, it consists of a weight attached to a long string hanging from the ceiling and set to swing back and forth like the innards of a grandfather clock. There are numerous Foucault pendulums around the world; the original still swings in the Panthéon of Paris.

Suppose an Australian-based Foucault pendulum originally moves along the north-south plane. Because of Earth's rotation, each time the pendulum moves north it lags behind Earth's eastward motion and inches a bit west. Then when it moves south it exceeds Earth's motion and goes slightly farther east. The result is that as Earth rotates, the pendulum's plane of motion slowly precesses (changes its angle) in a counterclockwise path around a circle. For the northern hemisphere, this process happens in reverse (because movement away from the equator is northerly rather than southerly), leading to clockwise precession. Therefore a proper Foucault pendulum is an excellent indicator of which hemisphere you are in.

An even quicker way of discerning your location, assuming the skies are free enough of haze and glare, is to look at the patterns of the stars. Depending on your latitude and the time of year, you would see distinct arrays of constellations. While in Australia, New Zealand, and many other parts of the southern hemisphere you would likely see the Southern Cross; in much of the northern hemisphere that constellation is impossible to view. Instead you might encounter Ursa Major (the Great Bear), part of which is also known as the Big Dipper or the Plough. This northern hemisphere constellation offers a convenient way of locating the North Star, the friendly companion of sailors seeking northerly passage. As old sea captains would tell you, a sailor flung far needs a star to tell where he is, yar.

If anyone could use a reliable means for finding his way through life, it would be Homer. Remarkably, while he's doing his Christmas shopping, he encounters one of the most ancient astronomical instruments, used in navigation and time telling. The curious thing about this device, which sets it apart from its predecessors, is that it talks. A loyal guide and a good conversationalist—what more can you ask for? Sorry, Santa's Little Helper, in that department you're outmoded.

20

If Astrolabes Could Talk

Many connoisseurs of classic television delight in the quirky series in which animals, inanimate objects, and the like miraculously acquire the ability to talk, and then proceed to outwit the astonished humans who encounter them. The classic—of course, of course—is Mister Ed, the talking horse, who is often neigh-saying the advice of his owner, Wilbur. Then there is *My Mother the Car*, about a chatty automobile that drives its owner to distraction. Its owner is actually its son, you see, due to a bizarre spin of the wheel of karma. That series was parked pretty quickly. And who could forget that crazy *Simpsons* spin-off (featured in the episode "The Simpsons Spin-off Showcase") in which Grandpa dies and haunts a "love-testing machine" in Moe's Tavern, dispensing advice to the lovelorn? It wasn't exactly a long-lasting spin-off; it lasted only about 1/100 of a season (less than ten minutes, more precisely). Still, who could forget it?

Yet in all the years of civilization (since the dawning of situation comedy at least), no one has developed a series based on a talking astrolabe. What is an astrolabe, you ask? For the ancients it was as useful as horses were for the nineteenth century, cars for the twentieth century, and love-testing machines are for today. A flat representation of the heavens, astrolabes were used for telling the time (particularly at night, but also as a portable sundial during the day), determining the calendar date, finding the height of an object against the sky, surveying land, and measuring latitude. Swiss army knives are positively inadequate by comparison.

Homer procures his talking astrolabe in one of the most selfish acts in television history. It makes the characters of *Dallas*, *Dynasty*, *Desperate Housewives*, and the *Sopranos* seem like good Samaritans in contrast, and puts Scrooge, Dr. Smith (from *Lost in Space*), and the Grinch to shame. In the episode "'Tis the Fifteen Season," Mr. Burns hands out Christmas presents to his workers and their families. Upon receiving a Joe DiMaggio baseball card as a present for Bart, Homer sells it to Comic Book Guy. The card is so valuable that Comic Book Guy gives Homer every last dollar in his cash register. Then, instead of using the money to buy gifts for his family, Homer spends almost all of it purchasing the talking astrolabe just for himself. There's barely enough left to buy a scrawny little Christmas tree. No wonder only a mindless machine will speak to him. Only later in the show, after watching "McGrew's Christmas Carol," does Homer recognize his own greed, repent, and become the very model of generosity—out-Flandering even Flanders.

Homer's device seems useful mainly for remembering celebrity birthdays, which it reports to him with tinny alacrity. Real astrolabes, in contrast, speak only through the utility and elegance of their design. Thus, if you are carrying one into a movie theater or a play you are more likely to attract gasps of envy than groans of annoyance. "Whoa, look at that awesome astrolabe," fellow theatergoers are likely to mutter. "And it has a silent feature. Maybe it's set on vibrate mode."

The term itself derives from the Greek word *astrolabos*, meaning "instrument that captures the stars." Though invented thousands of years ago in Greece and widely used in the Arab world, the device was perfected during the Middle Ages and developed into quite an intricate mechanism.

In 1391, Geoffrey Chaucer, author of *The Canterbury Tales*, wrote a famous treatise about astrolabes that is the oldest "technical manual" in the English language. Although Chaucer penned the following words for a boy named "Little Lewis," they could well have been spoken by Homer to Bart during a moment of fatherly affection (and after a few cans of Duff beer): "My sone, I aperceyve wel by certeyne evydences thyn abilite to lerne sciences touching nombres and proporciouns; and as wel considre I thy besy praier in special to lerne the tretys of the Astrelabie."[1] This means something like, "Boy, evidently you've learned some science and math, so you should be able to figure out this astrolabe," to which the boy may have replied something like, "Ay, caramba!"

Chaucer went on in his treatise to describe the function of an astrolabe from his time. A copy of such a device, dating back to 1326, is known as the Chaucer astrolabe. Representing the first-known European example, it resides in the medieval collection of the British Museum in London. It is truly extraordinary in the complexity of its design.

As Chaucer detailed, the device is a brass disk, little more than five inches in diameter, hanging vertically from a small ring. This plate is finely etched with detailed information about the Earth, the sky, and various times of the day. One side is marked with gradations representing the angles of a circle and the days and months of the year. This side could be used for mathematical calculations as well as astronomy. On the other side, the hours of the day and the twelve signs of the zodiac (various constellations from Capricorn to Sagittarius) are displayed. There is also a list of saints and the calendar dates they are celebrated, including three especially associated with England. Latitudes are marked for a variety of cities, including Oxford, Paris, Rome, Babylon, and Jerusalem.

Attached to the main plate is a smaller disk, called a rete, that can freely rotate to any position. Arranged in the shape of a Y, the rete contains pointers indicating where various stars are in the sky, including a dog-shaped indicator pointing to what is undoubtedly Santa's Little Helper's favorite astronomical body, Sirius, the dog star. By adjusting the rete, an astronomer can "tune" the astrolabe to any given latitude or time of the year. For example, if Professor Frink wished to figure out where Orion's belt would appear on St. Basil's Day over the skies of Cucamonga, he could twist the rete of an astrolabe to the proper place and figure it out in a jiffy. Great gravy, that's some handy gizmo!

For those who didn't like to carry around full-sized astrolabes, they also came in pocket-sized quadrants. These contained astronomical information compressed into an area one-quarter of the size—thus easily handheld, like the communicators in the *Star Trek* television series. Rumor has it that users were tempted to pick up the devices and shout, "Scotty, I think I'm in the Oxford quadrant, sometime in the preindustrial age of Earth. Historians are closing in. Beam me up immediately."

Astrolabes were also used for astrology, which, contrary to what Professor Frink claimed in "Future-Drama," has no known scientific validity. Nevertheless, members of the public, impressed by these devices' intricate workings, thought that they could be used for forecasting personal as well as astronomical futures. Throughout history, astronomical events have often been associated with good or ill fortune, depending on how they were interpreted. Of these, perhaps there was no more menacing symbol than the coming of a comet. The sighting of comets often brought enormous trepidation. In the case when Bart discovers his very own comet, this fear is certainly justified.

21

Cometary Cowabunga

For a small town in an unassuming state (hey what is that state, anyway?), Springfield sure has seen its share of disasters. From lethal radiation to alien invasions, it's witnessed them all. Irwin Allen, the producer of *The Poseidon Adventure*, *The Towering Inferno*, and numerous other disaster films, could have planted a web-cam on its streets and gathered enough scenes for his entire career. Fortunately, the Simpsons and their neighbors are a hardy bunch and seem to have held up well under the strain. Everything seems "okely-dokely," as Flanders often reminds us.

Perhaps one reason Ned thinks everything is okay is that he has his very own bomb shelter, just in case. Little does he know that when he needs the shelter the most, he won't be able to use it because it's blocked by his fellow townspeople. This happens on the day a comet heads right for the town.

The comet is first spotted by Bart, under most peculiar circumstances. Ordinarily, Bart is disinclined to spend much time observing nature—unless it is to trap crawly, slimy animals and release them at the most inopportune moments. Capturing the motion of celestial bodies, even if they ooze eerie streams of particles, just doesn't have the same gross-out factor. Rather, like Sisyphus of myth, who was cheeky enough to foil the plans of the gods and received eternal punishment for his impertinence, Bart is brought to the task as retribution. After Bart ruins a school weather balloon experiment by attaching a mocking caricature of the principal that unfolds as it launches, an irate Skinner forces the ten-year-old culprit to become his astronomical assistant. That's when Bart makes his earth-shattering (or at least Springfield-shattering) discovery.

Skinner has long wanted to find a comet to call his own. He claims that he discovered one once, but "Principal Kohoutek" beat him to reporting and naming it. It seems that Skinner is referring here to Comet Kohoutek, first sighted in 1973 by the Czech astronomer Lubos Kohoutek. Although I'm sure he's a man of great principle and you could say he's been a principal astronomer, Kohoutek has never actually led an American school. Regarding the third point, you could say that he and Skinner have much in common. Kohoutek has worked at various observatories, making numerous discoveries of comets and asteroids.

With a young, trustworthy assistant (or at least a potentially expellable indentured servant) by his side, Skinner hopes to etch his name into the hallowed annals of astronomical discovery—like Kohoutek, Alan Hale, David Levy, Carolyn and Gene Shoemaker, and so forth, who have each found numerous celestial bodies. For Bart, waking up at four-thirty in the morning is the cruelest part of the ordeal. Before assisting with Skinner's project, he didn't even know that ungodly hour existed. For comet-hunting astronomers, however, the nocturnal arena is the only playing field, and if you miss it you're out of the game.

Focusing his telescope on a seemingly empty patch of the sky, Skinner instructs his sleepy aide to take down notes about what he finds at various coordinates. As typical for astronomical measurements, he characterizes points in the sky through their right ascensions and declinations. These are used in the same way that longitude and latitude help us specify locations on Earth. Right ascension divides the celestial dome from east to west into hours of the clock. Just like the sun's daily motion, stars rise in the east and set in the west each night. Therefore, astronomers can distinguish the positions of stars by what times they ascend from the horizon. Such right ascension measurements are equivalent to using the hour of sunrise—indicated through a common standard such as Greenwich Mean Time—to figure out the longitude of someplace on Earth. What right ascension is to longitude, declination is to latitude. It tells astronomers how far north or south (above northerly or southerly points on Earth) an object is in the sky. Recorded in terms of compass angle, declination ranges from 90 degrees at the North Pole to –90 degrees at the South Pole—with the equator representing exactly 0 degrees. In the distant past astrolabes were used to make such determinations of celestial position, but today's telescopes have their own gauges, enabling Skinner to read off the coordinates and Bart to write them down.

Given the sad state of Springfield's sky as reported in "'Scuse Me while I Miss the Sky," it is not surprising that at first the principal and his boy-servant don't see anything of interest. Then Skinner spies the errant weather balloon with his caricature on it and runs after it to bring it down. If he can't find a comet, at least he can try to salvage what remains of his reputation. To that, Skinner's mother, Agnes, would probably retort that he shouldn't even bother—there's nothing left of it anyway.

While Skinner is off balloon-chasing, Bart makes the discovery of a lifetime. He spies a dirty snowball streaking through the heavens—in other words, his very own comet. After Bart promptly reports it to an observatory, it becomes known immediately as the Bart Simpson comet. The process by which it is named after him is unrealistically quick; normally objects such as comets and asteroids are reported to the Minor Planet Center at the Harvard-Smithsonian Observatory, where they are verified and tracked well before being given an official name.

The next day, Bart, fame in hand, is invited to join the "Super Friends," a group of brainy kids at school with nerdy nicknames such as Database and Report Card. Because of Bart's discovery, they dub him "Cosmos." While eating lunch together with his new companions, Cosmos mentions that his comet is visible out the window in broad daylight. The Super Friends rush to Professor Frink's observatory when they realize to their horror that it must be hurling at lightning speed toward Earth. In fact, as Frink determines, it's rushing directly toward the heart of Springfield—on a collision course with Moe's tavern, to be precise. Moe has had many icy arrows slicing through his heart before, but this one is inordinately cruel. He doesn't even get a "Dear Moe" letter to use as a tear- and beer-soaked souvenir placemat.

To save the town, Frink comes up with a plan. He proposes the launching of a rocket to intercept and blow up the astral invader. After firing the rocket, the townspeople are mortified when it misses the comet and instead pulverizes the only bridge out of town. Now, with no hope of escape and the comet due to arrive in six hours, they are really stuck.

That's when Flanders's bomb shelter comes in, as a haven of last resort. It's big enough to hold two families, which comes in handy when Homer pressures him to let the Simpsons in, too. Being a Good Samaritan, Flanders complies. But then all the neighborinos barge in, from Krusty to Barney, greatly overcrowding the shelter. Someone has to leave and face the fury of the icy intruder, but who? Homer rather ungraciously gives Flanders the boot, and he must bravely confront the comet outside alone. After a few minutes of rumination, Homer realizes his cruel mistake and decides to offer Flanders company. Soon everyone joins them, abandoning the shelter and courageously singing away the minutes until their likely doom.

The moment of truth arrives. The comet blazes for an instant then breaks up into gazillions of small pieces. Springfield's extra-thick, toxic atmosphere has pulverized the celestial marauder. Only one big piece remains that heads directly toward Flanders's bomb shelter, smashing into smithereens. The townspeople's solidarity with Ned has saved their lives.

Comets are objects of fascination and fear. Along with the sun and the planets and their moons, they form significant components of the solar system—essentially the material leftover from the formation of the larger spherical bodies. Following the laws of gravity, each follows an orbit around the center of the solar system. Yet, compared to planets and moons, comets tend to follow a much wider range of orbital patterns—far more stretched-out—spending the bulk of their time well beyond the range of Neptune, the outermost planet. (Pluto is now called a "dwarf planet," with something less than true planetary status.) Only briefly do they travel to the inner solar system, which is when we can best observe them. Because of their large orbital periods, they often seem to arrive like a bolt out of the blue. While some comets have well-known trajectories—Halley's comet being a famous example—the vast majority have yet to be tracked. Hence, they offer a great source of consternation; we never know when one will appear seemingly out of nowhere and come perilously close to Earth, maybe even colliding with our planet.

There are two key places where comets reside if they're not in our part of the solar system. The first, called the Kuiper Belt, lies just beyond Neptune's orbit and extends outward along the solar system's orbital plane. These have comparatively short orbits of less than two hundred years—visiting our part of space relatively often. Far less familiar are the comets populating the Oort Cloud, a spherical region trillions of miles in radius that surrounds the solar system's disk. This incredibly vast shell, spanning almost half the distance to the nearest stars, harbors approximately one trillion comets, each taking roughly one million years to orbit the sun. Occasionally, the gravitational tug of another star wrests one of these from its orbit, propelling it toward the inner solar system. Then astronomers, in the manner of Bart's discovery, report the presence of a new comet.

A tasty analogy illustrates this situation. Imagine the solar system's planetary orbits to be a donut resting in the middle of a plate on the Simpsons' kitchen counter. The outer ring of the donut is Neptune's orbit and the inner ring is Mercury's, with the other planets represented by the stuff in between. Now suppose that when the donut was dumped out of the bag, lots of extra crumbs spilled out. While some stuck to the donut, others scattered around the plate. These peripheral crumbs are the Kuiper Belt objects. Yet other crumbs are embedded in the gigantic mound of whipped cream that Homer has sprayed all over the donut. They comprise the Oort Cloud. When Homer lifts up the plate to move it to the table, some of these crumbs become dislodged, falling inward toward the donut, maybe even colliding with it. Homer gobbles down the donut, leaving the ort for Bart and Lisa. (Ort with one "o" means "leftover food.")

A long-standing popular myth imagines comets as fiery objects, sporting long burning tails like streaming fireworks. Actually comets are extremely cold, with their nucleus (central part) a cluster of dirt, rock, and ice some one to ten miles across. Their tails develop during the brief part of their journeys that they travel relatively close to the sun. Solar energy evaporates some of their ice, causing a release of vapor and dust. When light bounces off the trail of released particles, like reflections from a long sequined dress, we observe comets' tails. Comets also leave streams of ions, atoms from the evaporated gases that have their outer electrons wrested away by sunlight.

Close cousins of comets, with a similar range of sizes but a different composition, are asteroids. Asteroids are rocky bodies orbiting the sun within a range similar to that of the planets. Vast quantities of these occupy a zone between Jupiter and Mars called the asteroid belt. Others maintain closer orbits, even intersecting Earth's region, and in rare cases colliding with our planet. These are called Near-Earth Asteroids (NEAs). Together with short-period comets, they belong to a category called Near-Earth-Objects (NEOs), carefully tracked by astronomers because of their hazardous potential.

The main danger to Earth lies in NEOs ranging from about 150 feet to several miles in diameter, with the potential extent of their damage increasing with their size. If a comet or an asteroid the size of a large building collides with Earth, it can create a local catastrophe. Like a bomb, it generates enormous amounts of energy, wiping out whatever happens to lie near the point of impact—trees, houses, and so forth. In 1908 an asteroid or comet hit a forested region of Siberia called Tunguska, wholly obliterating vast tracts of forest. Although thousands of reindeer died in the blast, fortunately its remoteness spared people from being killed. If a similar explosion had happened in a major urban area such as Shanghai or Calcutta, tens of thousands could have perished.

Larger comets and asteroids, the size of villages or towns in girth, though less common in our region of space, represent far deadlier threats. The outcome of colliding with one of these titans would be absolutely terrifying. If an object two miles wide or bigger hit our planet, it would generate a blast of millions of megatons, forcing enough dust into the air to block out sunlight for months. This would lower Earth's temperature significantly, wiping out crops worldwide and potentially resulting in the extinction of numerous species. Many scientists believe that such mammoth cosmic collisions have occurred regularly throughout geological history, producing a fossil record of mass extinctions. Notably, an impact off the coast of Mexico sixty-five million years ago likely heralded the final gasps of the dinosaur age.

To try to reduce the chance of future collisions, astronomers have developed a worldwide tracking system, connected with a program called Spaceguard. In the United States a major center for observing NEOs is Arizona, where the Spacewatch survey at the University of Arizona and the LONEOS survey at Lowell Observatory cast a steady gaze on the heavens looking for astral intruders.

Astronomical missions around the globe have managed to identify more than two-thirds of the likely number of large NEOs.

So far, none of the NEOs tracked seem on a collision course with Earth. However, because new comets regularly emerge from the Oort Cloud, and older comets and asteroids can be diverted in their paths by other objects, the threat of cosmic impact remains a frightening fact of life. If a comet were just about to hit, there's nothing we could do. However, if astronomers determined that one would collide with Earth in the manner of years or decades, they could send up a craft to try and divert it. Blowing up the object would be unwise, because its center of mass, and many of its fragments, could still head toward Earth. However, an explosion that broke off a small piece of the body could divert the bulk of it enough that it could avoid our planet.

Although collisions with large asteroids and comets are fortunately rare, the Earth often passes through regions of space with smaller rocky objects. Much of the debris that falls to Earth burns up in the atmosphere, leading to spectacular meteor showers like the one observed by Lisa and others in the episode "'Scuse Me while I Miss the Sky." Rocky remnants that make it to the ground, called meteorites, are valued by scientists for what they tell us about the origins and composition of the solar system and for their potential to contain evidence of organic (carbon-based) chemicals from beyond Earth. These could point to the existence of extraterrestrial life. This question is tricky, however, because when a meteorite lands it is immediately invaded by all kinds of terrestrial organisms, masking its original conditions.

Thus, aside from major intruders that manage to break through, Earth's atmosphere is like a soft comfortable cushion, protecting us from the harsh realities of outer space. It shields us from certain types of lethal radiation, helps moderate temperatures around the globe by distributing heat, and reduces the amount of debris that falls to Earth. To see what the surface of Earth would be like with no atmosphere, look at the scarred visage of the moon, pockmarked with myriad craters.

Why would someone want to abandon that comfy cushion and venture into the cold, dark void beyond, where all manner of hazards abound? Homer asks that question to himself, no doubt, every time he lifts his posterior off his couch. Yet like the legendary Odysseus, whose chronicles were told by a different Homer (the ancient Greek poet), he has often left behind familiar comforts to face incredible perils. The Cyclops, for instance, had nothing on the combined venom of Patty and Selma. And the irresistible lure of the Sirens with their seductive songs were no match for the sweet whiffs of all the donut shops Homer must pass to get to work.

If Homer can face these perils and temptations with valor and resolution—and sometimes not even crying when one of the donut shops is closed—then surely he has the mettle to leave behind the cushiness of Earth and blast off into space. That's one small step off the sofa and one giant leap into the colossal vacuum. Woo hoo!

22

Homer's Space Odyssey

Humankind has long sought to reach out to the stars. Shedding Earth's protective cocoon and heading into the vast interplanetary void represents one of the supreme goals of our race. Our spirits strive to lift us higher—even as our bodily limitations render space travel a formidable challenge.

During the past half-century a hardy group of pioneers have braved the rigors of the harsh environment beyond Earth's atmosphere. These individuals have gone through intense training to learn how to cope with conditions ranging from zero gravity to stomach-wrenching acceleration, from cramped quarters to the unimaginable vastness of space, and from absolute silence to the rocking sounds of Sonny and Cher. Whoa, that's intense.

The rigors of a space journey are such that virtually every moment must be carefully planned, from the precise time of launch and the instant when booster rockets must be fired, to the type of music waking the astronauts each day. Recently, Paul McCartney serenaded the astronauts in the International Space Station through a live hookup, and that took almost five decades of planning—starting with him buying a bass guitar and performing at the Cavern Club and other Liverpool locales; embarking on careers with the Beatles, Wings, and solo; appearing on *MTV Unplugged* and *The Simpsons*; receiving a knighthood; and finally making time in his busy schedule for the NASA gig. One false career move and perhaps Davy Jones of the Monkees would have had to step in, lest the musical aspect of the mission be jeopardized. As we can see, truly every aspect of our bold adventure in space must be meticulously prepared.

In the episode "Deep Space Homer," we witness the next hero to step into the shoes of John Glenn, Yuri Gagarin, Neil Armstrong, and Sally Ride. It's an inanimate carbon rod (ICR), bringing Homer along for the adventure of his life.

Homer's journey begins when a rod from the nuclear power plant beats him for "worker of the year." Everybody laughs at him, even his family. Depressed, he turns on the TV only to see a boring, poorly rated show about space. When he complains to NASA, its public relations staff discovers an opportunity to boost the ratings of its broadcasts. They invite Homer, representing the "average blue-collar American," to participate in its next space mission.

With Buzz Aldrin, the second man on the moon, as one of his crewmates, Homer blasts off into space but makes the mistake of opening a packet of potato chips he has smuggled aboard. Because of the zero gravity conditions, the chips follow their natural inertial paths and scatter throughout the ship. As Newton pointed out, in situations where external forces (such as gravity) have no effect or balance out, objects proceed along straight-line paths at constant velocities. The only way they can be slowed down or stopped is by introducing an extra force. For example, a steel screwdriver floating off into space could be stopped by a magnet. To gather up the potato chips, therefore, the crew needs to act decisively. Otherwise, they'll keep going forever—or at least until they muck something up—and the crew might as well say good-bye, Mr. Chips, and good-bye, Mr. Ship.

In a Newtonian moment, Homer determines that his mouth could provide the necessary force to capture the potato chips. In a parody of a scene from Stanley Kubrick's monumental film *2001: A Space Odyssey*, he tries to eat up all the chips while floating around the craft. In the process, he manages to shatter an ant colony that had been brought on as an experiment to see if ants could be taught to sort tiny screws in space. Released from their case, the ants invade the ship, short out its navigational circuits, and further endanger the mission.

In the midst of this mayhem, the crewmembers are treated to a live broadcast of the mellow sounds of James Taylor. Procuring Taylor proves fortuitous indeed. While serenading them with his laid-back hit songs "You've Got a Friend" and "Fire and Rain," Taylor learns of their plight and recounts a similar ant infestation he once had at his summer cottage. Balladeer Art Garfunkel solved the problem by creating vacuum conditions outside the door, which sucked the ants outside. Why not do the same thing in space?

The NASA scientists decide to take Taylor's advice. After putting on their spacesuits, Aldrin and the crew blow open the hatch, and the ants are ejected. The trouble is, Homer forgets to secure himself and almost flies away from the ship. While pulling on the opened hatch, he breaks off its handle. Now the hatch can't be closed during reentry. To secure it, Homer jimmies it shut with an identical copy of ICR that he's managed to find on the ship. Thanks to Homer's snappy solution, the crew returns safely to Earth. Homer is incensed when ICR receives a hero's welcome, complete with a parade, and his own efforts are once again ignored.

"Deep Space Homer" well captures the efforts by NASA, the European Space Agency (ESA), and the Russian Federal Space Agency (RFSA) to diversify their missions by identifying people of nontraditional backgrounds and training them to be crew members.

For example, NASA's "Teachers in Space" program was designed to bring educators on board who would return to their schools and teach children about life in space. Tragically, this program was set back due to the 1986 space shuttle *Challenger* disaster in which teacher/astronaut Christa McAuliffe and six other crew members died. Because of that tragedy and the space shuttle *Columbia* disaster in 2003, another teacher/astronaut, Barbara Morgan, who trained with McAuliffe, was grounded for many years. In 2007, however, NASA plans to send her on an assembly mission to the International Space Station, helping to educate a new generation of children about spaceflight.

In conjunction with a private agency called Space Adventures, the RFSA has taken a different approach. They have opened up some of their missions to space tourism, offering wealthy civilians opportunities for spaceflight if they are willing to pay millions of dollars for the privilege. The first space tourist was Dennis Tito, a California businessman who was sixty years old in 2001 when he spent $20 million for a weeklong visit to the International Space Station. He traveled to the station aboard a Russian Soyuz rocket along with several trained cosmonauts. Two other space tourists followed: Mark Shuttleworth in 2003 and Greg Olson in 2005. While these flights have helped raise funds for the Russian space program, NASA initially expressed opposition to the program, fearing the risks of inadequately trained civilian passengers. With the success of the program and the publicity it has generated, however, NASA's opposition has faded. After all, none of the passengers so far has opened potato chip packets, released ant colonies, or performed James Taylor songs.

In 2006, the-forty-year-old Iranian American businesswoman Anoushe Ansari become the first female space tourist. Packed into a cramped Soyuz TMA-9 capsule, along with a U.S. astronaut and a Russian cosmonaut, they blasted off from the Baikonur Cosmodrome in Kazakhstan and rendezvoused with the International Space Station soon thereafter. Ansari spent eleven days aboard the station before returning to Earth.

The Ansari family has close connections to space tourism. They funded a $10 million award, now known as the Ansari X Prize, to the first private organization able to launch a human-occupied spacecraft into space twice within a two-week period. The 2004 prizewinner, Burt Rutan of Scaled Composites, is an American businessman who specializes in innovative aircraft. He used his inventive skill to design SpaceShipOne, a prototype for private reusable space vehicles.

In 2005, Rutan teamed up with Richard Branson of the Virgin Group to develop a fleet of private spacecraft, based on Rutan's archetype. Branson's new company, Virgin Galactic, is planning to inaugurate lower-cost ($200,000 per passenger) commercial spaceflight in 2008. Then, space tourism will no longer be limited to the Montgomery Burnses of this world, but will also be available to less wealthy passengers—those willing to spend hundreds of thousands of dollars on flights, that is. So if someone like Marge was choosing between spending $1,000 on a round-trip flight to Las Vegas and $199,000 on gambling, or just $200,000 for a flight into the void, she might avoid the moral dilemma and decide to spend it on spaceflight. Homer would just need to work overtime for a few millennia to pay off the bill.

Once the less-than-rich are able to afford space travel, imagine all the fabulous leisure possibilities. Eventually, even the extraordinary could become mundane. If spaceflight became commonplace, it might be considered just another computer-scheduled component of travel itineraries. Customers booking formerly direct flights from San Francisco to Los Angeles might be stuck with an overnight layover near the Van Allen belt. In the future, when booking flights online, it would be wise to specify "nonstop, terrestrial only," if you don't want to find yourself floating in the cabin.

We might imagine custom-made space excursions with virtually unlimited themed possibilities. On *The Simpsons*, there would be ample material for possible future spaceflight scenarios. Krusty could set up kids' birthday flights, and any stomach-churning could be chalked up to the high *g*-forces, not the food. Eleanor Abernathy, the Crazy Cat Lady, could advertise "Feline Fancy Flights," similar to the film *Snakes on a Plane*, but with purring animals hurled at passengers instead. Comic Book Guy could pilot a replica of *Star Trek*'s *Enterprise*—paying Groundskeeper Willie to serve as his chief engineer—and erase the shame of its "worst episode ever."

If doomsayers turn out to be right, civilians ought to get accustomed to space travel. Someday Earth could be threatened by a calamity deadly enough to destroy our civilization. Suppose, for example, astronomers discover a colossal comet or asteroid heading straight for our planet that couldn't possibly be diverted in time. Then evacuating Earth and establishing space colonies elsewhere could represent a viable option.

In one of the Treehouse of Horror episodes, the residents of Springfield face such an emergency evacuation. Homer and Bart manage to escape Earth just in time to avoid its hour of doom. Unfortunately, they end up on the wrong spaceship.

23

Could This Really Be the End?

For everything there is a season. In television series and human lives, there are times of growth and times of decay. Sometimes endings are sudden; other times hopeless situations drag on and on. The original *Star Trek* series was canceled after only three years, yet it has lived on in four spin-off series, as well as numerous movies, books, and other media. Human limitations being what they are, however, a steady infusion of younger actors has been needed to keep the franchise fresh and vibrant. It just wouldn't do to have the original cast members still fist-fighting aliens while in their seventies.

With the distinct advantage of being animated, *The Simpsons* has thus far avoided such perils. Nevertheless, pundits wonder if the appearance of *The Simpsons Movie* has signaled that the end of the television series is in sight. I hope not. I would like my grandchildren, great-grandchildren, and so forth to be able to watch new episodes of the show. But alas, someday a decision will be made to terminate the series. Might there then be sequel films and spin-offs? (The episode "The Simpsons Spin-off Showcase" offers tongue-in-cheek hints of possible postseries plans, including a "Simpsons Smile-Time Variety Hour" with another actress replacing the original Lisa.)

What about life on our side of the screen? Human civilization, despite numerous setbacks and countless cast changes, has managed to survive for many thousands of years on Earth. We hope it will thrive for many more seasons. Yet we must face the terrifying prospects that it could someday be "cancelled" through natural or other means. Could there be "spin-offs"—sequels to our culture, perhaps on other planets?

The catastrophic result of a large comet or asteroid is but one of the many possible calamities that could someday imperil civilized life on our planet. We have seen how such a blast could propel trillions of tons of dust into the air, blotting out the sun for months, drastically lowering temperatures around the globe, and causing mass extinctions, as in the last days of the dinosaurs. In some cases the threatening body could be diverted, but only if there was enough time. Otherwise we'd be doomed.

A similar global deep freeze would result from large-scale nuclear war. Even with the end of the Cold War, global nuclear conflict remains a formidable risk. Who knows when another arms race could ensue? Thousands of missiles, if launched, would not only bring radioactive cargo, they'd also generate enough of a dust cloud to trigger a long era of frigid darkness known as nuclear winter. Food sources would be wiped out and advanced life on Earth could be vanquished forever.

In one of the few *Simpsons* episodes that truly seems dated, the 1999 Halloween segment "Life's a Glitch, Then You Die," another global threat is addressed: the possibility of universal computer malfunction. Comparing the effects of computer error with that of asteroid impact or nuclear catastrophe is like equating the discomforts of the common cold with the ravages of the bubonic plague. Nevertheless, during the late 1990s, the "Y2K bug," a computer malfunction associated with the year 2000, came to be seen by some experts as not just a nuisance but as the potential trigger of worldwide apocalypse.

The Y2K problem concerned computers equipped with date functions that did not take into account years beyond the 1900s. These limited date functions were introduced as a way to save memory—they stored only the last two digits of the year, not the whole thing. Therefore, if they weren't updated, the year 2000 would not be recognized and the electronic calendars would return to 1900. As a consequence, the flawed computers would treat backup files from 1999 or earlier as if they were more recent than files from 2000 (erroneously dated 1900). They would erase the newer versions in favor of the old, or perhaps even wipe out all the files. This would cause mayhem to bank accounts, government files, and other records, generating total havoc, so it was supposed. To ward off possible disaster, billions of dollars were spent around the world to update computer software to take into account the new millennium. Also, essential systems everywhere were backed up.

Many businesses and agencies designated a "Y2K compliance officer" to carefully check all the computer systems to protect against failure. That individual needed to be technologically savvy and highly responsible. The fate of gigabytes of data, representing the records of numerous individuals, was riding on his or her shoulders.

In the "Life's a Glitch" episode, the Springfield nuclear plant picks Homer for this duty, and you can guess how effective a job he does. When January 1, 2000, comes around and the famous New Year's countdown at Times Square takes place, not only does the Springfield plant fail, it sets off a chain reaction that causes global pandemonium. Traffic lights go on the blink, a revolving restaurant spins out of control, airplanes fall from the sky, and even household appliances fail to function. Nothing with electronics seems immune, from waffle irons to refrigerator icemakers. When Homer tries to open a carton of milk for his late-night snack, it sprays out in all directions, presumably because of an imbedded computer chip gone haywire. The widespread failure of technology rapidly leads to social breakdown. Amid scenes of large-scale looting and mass panic, Reverend Lovejoy declares that the Day of Judgment has arrived.

While fleeing the mayhem, the Simpsons encounter a very uncomfortable Krusty the Clown. The Y2K disaster has tripped his pacemaker into high-speed "hummingbird mode." After flapping his arms for some time, he collapses on the ground. A saddened Bart discovers a note in Krusty's pocket with an invitation to Operation Exodus, a plan to evacuate the Earth. The Simpsons realize that the letter offers them a ticket to salvation. Clutching it carefully, they run toward an awaiting rocket ship, anticipating a new tomorrow in space.

Standing in front of the spaceship is a guard whose job is to let in only the best and the brightest—Bill Gates and Stephen Hawking, for example. Reading from a list, the guard announces that Lisa is permitted on the vehicle and is allowed to bring along only one parent. Without hesitation she chooses her mother. Lisa, Marge, and Maggie board the ship, leaving an angry Homer and Bart in the lurch. They manage to find another rocket ship, filled with those left behind by the first vehicle—in other words, those deemed superfluous. While the first rocket with the female Simpsons and other notables is heading toward a new life on Mars, the second ship with the male Simpsons and other expendables turns out to be aimed straight for the sun. When those around them start singing, Homer and Bart decide to hasten their demise by ejecting themselves into the vacuum.

In real life, the Y2K crisis never represented the apocalyptic scenario some had feared. Through preparations costing billions of dollars in total, software updates and backups headed off any significant problems. Perhaps due to this careful planning, as 2000 began, the moment after the ball at Times Square dropped was scarcely different from the moment before. Even if the worst had come to pass, it's doubtful that the computer failures would have affected many people, save the inconvenience of trying to reconstruct records that had inadvertently been erased. It could have been a major headache, but not doomsday.

If we want to be as gloomy as Hans Moleman and ponder potential apocalyptic scenarios, unfortunately there are far worse things that could transpire. Global warming, if it continues unabated, could produce major ecological catastrophes. Major portions of Earth could be rendered desert. The Gulf Stream could shift and the coast of northern Europe lose its protection against the Arctic cold. Pollution and overdevelopment could continue to eradicate untold species, disrupting the food chain. Conceivably, at some point our environment could be unfit for civilization as we know it.

If, in the future, the human race faced a strong possibility of extinction from impending disaster, establishing space colonies may well prove the only hope for our species. The feasibility of large-scale evacuation would well depend on how extensive spaceflight is at the time. The current shuttle program would clearly be inadequate to transport millions of people up into orbiting stations and then on to extraterrestrial sanctuary. Perhaps something like a space elevator would be more effective. Researchers have proposed attaching pencil-thin ribbons tens of thousands of miles long from Earth up to counterweights placed in geosynchronous orbit. Geosynchronous means keeping exact pace with the Earth's spin, and therefore always above the same place on the globe. If the ribbons were sturdy enough (constructed, for example, of the ultra-strong, super-thin chains of molecules known as carbon nanotubes), gravity and Earth's rotation would act in tandem to keep them permanently stretched out. The ribbons would serve as cables for space elevators that would transport material up through the atmosphere and out into the blackness beyond.

The Ansari family, the Spaceward Foundation, and NASA supported a 2006 X Prize competition in Las Cruces, New Mexico, called the Space Elevator Games, targeted at challenging research teams to develop prototypes for durable but lightweight ribbons and vehicles. The goal was to encourage the construction of a space elevator by 2010. The best entry for the "power beam" race, developed and assembled by the University of Saskatchewan Space Design Team, was a platform that climbed a 200-foot-long tether in 57.5 seconds—a new speed record but a hair short of being fast enough to claim a $200,000 prize. Undoubtedly, the team is preparing to set new records in future versions of the competition.

If an efficient enough space elevator were constructed, it would greatly aid in evacuating Earth in the event of impending apocalypse. Evacuees would step into cabins on the lift, called "climbers," that would slide up the ribbons and then off into space, docking with space stations. There they could board enormous space arks that would take them to other worlds.

At that point, the question would be where to go. Perhaps the colonists would settle in enclosed communities on the moon. However, unless conditions on Earth were absolutely untenable, it's hard to imagine life on the moon being more pleasant. Somehow breathable air would need to be generated and sufficient water supplied for drinking and crops. There is some chance that the moon's craters in its unlit polar regions contain scattered ice crystals brought there over the eons through the bombardment of comets. These crystals could be spread out in thin layers or mixed with lunar soil, making extraction of the water a difficult chore. If the water and lunar minerals could support plants—brought there, perhaps, from Earth's deserts to insure that they are used to dry conditions—the plants, in turn, would produce oxygen to support humans and any other animal life in the colony. This ecosystem would need to be exceptionally fine-tuned, allowing little room for waste.

A better long-term solution would be to establish Earthlike conditions on the surface of another planet, most likely Mars. Aside from Earth, Mars is the planet in our solar system with the most favorable conditions. The outer four worlds, such as Jupiter and Saturn, are enormous balls of gas with crushing atmospheric pressure.

They may not even have surfaces to land on. The inner two planets are scarcely better. Mercury, being so close to the sun, has a daytime temperature that is far too hot. Venus, though similar in size to Earth, has a thick poisonous atmosphere with a runaway greenhouse effect. Dense clouds trap heat, making the surface searing. If colonists of the future wanted to escape the effects of global warming, Venus would *not* be the place to go. At least Mars has reasonable temperatures, a solid surface, and gravity that, though weaker than Earth's, settlers could easily adjust to. True, with its thin atmosphere, gale-force winds, and lack of liquid water, it's no paradise, but perhaps with some engineering it could be made more like home. Compared to Venus or Mercury, Mars's conditions are bearable.

Terraforming, the process of transforming a planet into a near-replica of Earth, is a highly controversial subject, pitting advocates of human space colonization against those who advocate preserving native conditions at all costs. In some ways, these echo arguments about earthly development versus preservation. If you make a desert bloom through irrigation, it's no longer a desert. If you knock down a run-down historical district and replace it with efficient modern housing, it's no longer historical. The benefits and costs must be carefully balanced.

For example, suppose a group of snobbish investors from Shelbyville wanted to remake Springfield into a showcase community. They could buy up its land, raze Moe's tavern, sell the schools to private consortiums, and replace the nuclear power plant with an ultra-efficient high-tech wind-generation unit. The streams could be purified and restocked with ordinary, two-eyed fish. All the incompetent workers and technicians could be replaced with experts. Fancy shops with high-priced European goods could occupy a gleaming galleria on the site of the old Kwik-E-Mart. New theaters and symphony halls could be constructed. (Springfield tried to build the latter once, but attendance was pitiful and it folded; this time cultured audience members could be bused in.) In short, Springfield could become a model of urban sophistication. But where would be the charm? Where would be the history? What would happen to its hapless displaced workers? And what would become of poor Blinky, the three-eyed fish? Would he be forced to spend his waning years in some run-down aquarium on the outskirts of town instead of roaming free in glowing green waters?

Terraforming would involve a similar trade-off. If Mars, for instance, were remade to resemble Earth, then its original landscape would be lost and any native life forms—assuming they exist in some unexplored niches—would be potentially wiped out. Life on Mars has yet to be found, but considering that living organisms on Earth thrive under extreme conditions (such as extremophile microbes living in dark crevices deep underground, drawing energy from chemical processes), astronomers still hope it exists somewhere. A radical transformation of the Martian environment might reduce that chance to zero.

On the other hand, if Earth were no longer a safe haven, or if it someday became too overcrowded, terraforming Mars might be the only realistic option, especially if interstellar travel had yet to be developed. A number of researchers—such as the aerospace engineer Robert Zubrin of Pioneer Astronautics, the astrobiologist Christopher McKay of NASA Ames Laboratory, and the British dentist-turned-science-writer Martyn Fogg—have advanced various proposals for ways of making the red planet homier. These ideas include placing mirrors more than a hundred miles across in orbit above the Martian polar ice caps to reflect sunlight onto the surface, vaporize its carbon dioxide, and generate a thicker atmosphere, and manufacturing halocarbon gases to trap energy and create a greenhouse effect. This would raise the temperature above the freezing point of water for much of the Martian year, allowing rivers and streams to flow along the surface, supporting plant and animal life. Many scientists believe that liquid water once shaped the Martian terrain; perhaps it will flow again someday.

The transformation of Mars into a fully habitable planet would be a gradual process, taking many centuries. While at first the atmosphere would be unbreathable, requiring settlers to wear spacesuits all the time or to live in enclosed domes, eventually the plants would convert enough of the carbon dioxide into oxygen to enable people to breathe freely outside. Conditions might never be quite as pleasant as Earth's, but at least the human chronicle would continue.

It's easy to imagine a strong, determined woman like Marge as one of the first Martian settlers. She has proven herself handy with tools, building all manner of structures and furniture in the episode "Please Homer Don't Hammer 'Em." In "Strong Arms of the Ma," she becomes a weightlifter and shows how powerful she can be. Moreover, despite having an inept husband, she's been able to raise and protect a family and keep a household well managed. She strives to be honest and fair-minded and rarely loses her balanced perspective. As Jebediah Springfield would say, her noble spirit would "embiggen" the humblest Martian colonist. Who could ask for a more suitable pioneer?

So if another massive computer bug, comet, nuclear disaster, or other apocalyptic scenario plagues the town of Springfield, a mission to Mars headed by Marge with Lisa as chief science officer would be most appropriate. When survival is at stake, conquering another planet makes a lot of sense. But what if the captured planet is our own and the conquerors are an alien race from a world orbiting a distant sun? Would we be so keen on planetary transformation if slobbering extraterrestrials are the ones trying to remake us?

24

Foolish Earthlings

For two decades, images of the Simpsons have been broadcast steadily into space. By now, the antics of Homer and his family have likely reached a number of planets within twenty light years of Earth. The three planets near the red dwarf star Gleise 876, for example, approximately fifteen light years away, could have caught the first season several years ago, not too long after it was first released on DVD. If cognizant beings are on any of those worlds with the ability to discern and decipher radio and television transmissions, they could have laughed, cried, or snorted along with the shows. So why hasn't the human race gotten back any fan letters? It wouldn't hurt for extraterrestrials to send us at least a little note.

True, a response to any of our broadcasts would take the same amount of time to reach us as our transmissions do to reach them.

Consequently, we may need to be patient about hearing what real aliens think of the show's depiction of extraterrestrial life. Yet there were other television series about aliens that are much older: *My Favorite Martian*, *Mork and Mindy*, and so forth, and countless other television programs of all sorts beamed into space for more than sixty years, enough time to spread out much farther than the *Simpsons* signals. Orson Welles's famous radio broadcast of *War of the Worlds* that caused such mass panic happened way back in 1938. In the intervening seven decades, the signals could have reached planets more than thirty light years away. If a civilization was advanced enough to detect these signals and determine that they came from an intelligent source, they could have responded by now. Yet we've heard nothing.

Since the 1960s, the Search for Extraterrestrial Intelligence (SETI) program has scanned the skies for radio signals conveying messages from alien civilizations, using telescopes such as the giant radio dish in Arecibo, Puerto Rico. Yet despite decades of trying to discern patterns among the noise indicating advanced communications, nary a meaningful utterance has been found—not even an interstellar "D'oh!" At various times there have been unidentified flying object (UFO) sightings, which a certain segment of the population has put forth as evidence that extraterrestrials are already here. Even former U.S. president and Nobel peace prize recipient Jimmy Carter once reported seeing a UFO. However, many of these events have later been explained as meteorological phenomena, weather balloons, experimental military aircraft, and so forth. In short, for all the extensive searching and reports of strange sightings, there has been absolutely no scientific evidence that extraterrestrial intelligence exists. Given the vast number of stars and planets in the Milky Way, with statistically at least a certain percentage having the necessary conditions for life, this lack of contact is surprising. As the great Italian American physicist Enrico Fermi once famously inquired, "Where is everybody?"

A number of researchers have attempted to offer intuitive solutions to Fermi's dilemma. Several notable scientists, such as Michael

H. Hart of the National Center for Atmospheric Research and Frank Tipler of Tulane University, have asserted that we must be the only habitants of our galaxy with an advanced civilization. According to Hart, with radio communication such a straightforward process, if intelligent aliens were out there somewhere they would have figured it out by now and sent signals. Tipler goes further, suggesting that any intelligent beings could have and would have conquered the galaxy by now, perhaps through robot surrogates that reproduce themselves and spread out throughout the stars like armies of drones. Hence, in this dog-eat-dog universe, no other mutts have marked their territory, so it's all ours.

Other scientists have been far more optimistic about the prospects for other cognizant beings in space. The late astronomer Carl Sagan, for example, argued that although the vast voids of space would make direct interstellar contact challenging, it's only a matter of time before it happens. His novel *Contact*, with its wormhole transportation scenario, embodied his deep hope that the formidable interstellar chasms could be bridged and friendly ties established with possible civilizations numerous light years away. It answered Fermi's query with a call for patience and determination.

Fermi's question need not be asked on *The Simpsons*, since the answer is obvious. Alien visits to Springfield have been well documented in the annual Treehouse of Horror Halloween episodes. From the second season onward, the one-eyed slobbering creatures Kang and Kodos, later revealed to be brother and sister, have been featured in title roles and bit parts. Though their haughty disdain for earthlings is clear and their desire to kidnap or conquer hapless humans remains a constant, their precise motivation is often vague. Do they view us as dangerous rivals, delectable treats, or harmless dimwits needed to be taken under their tentacles and tutored? Perhaps this ambiguity echoes the similarly mixed attitudes people have toward lower animal species, seeing them in various contexts as hazards, food sources, or potentially trainable pets.

In the segment "Hungry Are the Damned," Kang and Kodos mark their first appearance on the show by whisking away the

Simpson family in a classic flying saucer. From the moment Homer and his kin are on board, the aliens begin to feed them as much as they can possibly eat. While the others gorge themselves and express much appreciation for the aliens' hospitality, Lisa begins to suspect that their drooling hosts have a sinister ulterior motive. She learns that once they arrive on the aliens' planet Rigel IV, they'll be guests of honor at a sumptuous banquet for which their hosts are saving their appetites. Her suspicions grow when she sees a tentacled beast preparing a pot, hunting for the proper spices, and reading a cookbook that appears to be titled "How to Cook Humans." Alarmed, she grabs the cookbook and runs to her family, explaining they are doomed to be the main course. As the family protests, Kodos blows dust off the jacket, revealing the title "How to Cook For Humans." Lisa blows off more dust, making the title appear to read "How to Cook Forty Humans." Finally, Kodos removes the last of the dust, revealing the cookbook's true title, "How to Cook for Forty Humans." The aliens chastise the Simpsons for their false suspicions and explain that if they were more trusting they could have experienced paradise. Lisa humbly concedes that she was mistaken about the aliens' intentions.

Later episodes reveal assorted plots by Kang and Kodos to take over Earth, although it is often unclear why they would even bother given their vastly superior powers. For instance, in "Citizen Kang," part of Treehouse of Horror VII, the two creatures kidnap two prominent U.S. politicians—President Bill Clinton and Senator Bob Dole—who vied with each other in the 1996 presidential race. Kang and Kodos, assuming the candidates' forms, run for president themselves. After Homer reveals to the public that they are aliens in disguise, Kang and Kodos explain that the American two-party system compels U.S. citizens to pick one of the two. The voters choose Kang, who forces human slaves—including the Simpsons—to build an enormous ray-gun to be aimed at an unnamed planet.

Treehouse of Horror XVII includes the segment "The Day the Earth Looked Stupid." Although its title references the classic 1950s movie *The Day the Earth Stood Still*, its theme of alien invasion is based largely on *War of the Worlds*. The first part of the segment imagines how a 1938 version of Springfield reacts in panic to Orson Welles's famous radio broadcast of that story. This leads to general skepticism, craftily exploited by Kang and Kodos, who launch a real invasion. Although their initial attack is swift, the resistance and occupation drag on for many years. Kang and Kodos finally explain that they needed to invade Earth because humans were working on "weapons of mass disintegration."

Although the motivations of Kang and Kodos are often unclear, at least we share with them a common tongue. By sheer coincidence, the Rigelian language they speak is identical to English. Linguistic confusion arises mainly when the phrases they use are imprecise. Thus, when in "Hungry Are the Damned" Kodos uses the expression "chew the fat," ambiguity about whether she plans to chat or chomp arises from a known distinction between the phrase's literal and figurative meanings in English. Such misunderstanding due to plays on words—abundant in comedies by Shakespeare, Oscar Wilde, and many others—is relatively mild compared to the true difficulties that would likely arise if humanity encountered a real extraterrestrial race.

Far more realistically, extraterrestrials would almost certainly communicate through languages with virtually no common basis with known idioms. Just as terrestrial languages have been shaped by the unique experiences of various peoples on Earth, extraterrestrial communication would be molded by each alien race's anatomy, history, and living conditions. Hence, any meaningful dialogue with extraterrestrial beings would likely require surmounting tremendous communication barriers.

In 1953, the writer G. R. Shipman published a piece titled "How to Talk to a Martian," envisioning the process by which linguistic anthropologists could decipher alien languages. Decrying stories imagining that extraterrestrials could learn Earth languages instantly through translation devices or telepathy, he called on other writers to investigate actual methods used to unravel unknown tongues. Shipman explained how linguists work with human informants speaking unfamiliar languages to identify common points of reference that can be used as stepping-stones toward full mutual understanding. He envisaged that the same techniques could be applied to extraterrestrial languages. "If the inhabitants of other planets use speech sounds as we do," he emphasized, "their language should yield to analysis by our methods as easily as any Earth language. The same would be true if they use any combination of other types of visual, audible, or tactile signals."[1]

If only it were that simple. Human language, as Noam Chomsky and others have pointed out, is acquired though dedicated brain functions that determine the process by which the grammar of each specific language is constructed. Thus all known languages are fundamentally shaped by a biological component produced through human evolution. Because we cannot assume that the evolution of extraterrestrial beings proceeded along similar lines, there is no reason to think alien communication will have anything like the underlying grammatical patterns we associate with human language. In other words, not only would it be extremely improbable that a real-life Kang and Kodos would speak English, it would also be highly unlikely that the structure of their grammar would have anything in common with known languages.

How then could we possibly fathom the musings of our counterparts from other worlds? In the SETI program, much hope for intercepting and interpreting alien communication lies in finding signals with measurable features based on universal mathematical or physical properties. For example, in 1959, Giuseppe Cocconi and Philip Morrison of Cornell University suggested in their influential article "Searching for Interstellar Communications" that a frequency called the "hydrogen line" (1,420 Megahertz or 1,420 million wave vibrations per second) would be a promising spot on the radio spectrum to hunt for extraterrestrial signals.[2] The hydrogen line is a readily observable radio emission frequency associated with neutral hydrogen that is commonly used as an astronomical benchmark. It was first detected by the Harvard researchers Harold Ewen and Edward Purcell in 1951. In 1960, Frank Drake of the

National Radio Astronomy Observatory in Green Bank, West Virginia, initiated Project Ozma, the first project to search for signs of extraterrestrial intelligence, homing in specifically on the hydrogen line. Since then, a number of other SETI missions have centered on the same region of the spectrum.

The idea is that advanced beings, even if they possessed wholly different physiologies and brain functioning, would still know that hydrogen is the most basic element, that it is common throughout the universe, and that it has certain distinct spectral lines. Moreover, the aliens would realize that the frequency region around the hydrogen line is "radio quiet": that is, relatively free of noise from other effects. They would also infer that other intelligent beings in the universe would reach similar conclusions. Therefore, they would see the hydrogen line as prime broadcasting turf.

As SETI researchers have pointed out, in content too, alien species might try to include references to mathematical or scientific commonalities. For instance, they might send pulses spaced according to the prime numbers, the Fibonacci sequence (each number in the sequence is the sum of the previous two), the digits of pi, or other fundamental patterns. Unless the beings have ten fingers, these would probably not be sent in decimal notation, but rather might be transmitted in binary form (0s and 1s) or yet another numerical system. SETI scientists have played a kind of guessing game trying to analyze signals for the wide range of possibilities.

In the decades of searching there have been only a few events that have set hearts racing with anticipation of possible discovery. One such incident occurred in 1977, when Jerry Ehman, volunteering at the Big Ear Radio Observatory, discovered a signal in the hydrogen line region far stronger than background noise. It was like standing in a silent cave and suddenly hearing a scream; you'd assume that someone else was there. Ehman was so surprised that he wrote "Wow!" on the page; hence, it's come to be known as the "Wow! signal." In all the years since then, however, nobody has been able to reproduce that strange cry in the dark. Hence, it was either an extraterrestrial race that only broadcast for a brief interval, or, more likely, a signal from Earth that somehow bounced back (perhaps off some form of space debris, as Ehman has suggested) and interfered with the observations.

Suppose someday we receive messages from a distant alien civilization. Would we and the other race have the capability of physically meeting each other or would we be doomed to a long-distance relationship? If the interactions between Kang and Kodos and the residents of Springfield are a gauge, perhaps remote contact is the way to go. On the other hand, if the right alien race came along, with attractive enough prospects for a warm relationship, many of us, like Moe, might hope that a friendly message would be followed by a more intimate rendezvous. I mean a meeting of minds, of course. The question then would be our place or theirs and—if we're the ones doing the traveling—how to bridge the inordinate distance between our species.

Given that we still haven't set foot on any other actual planet (unless you count the moon, which is considered a satellite, not a planet), interstellar travel likely is a long way off. Yet, less than a century ago, any form of space travel lay exclusively in the realm of science fiction. We are a persistent lot and, through our power of inventiveness, have found ways around numerous other technological impasses. Consequently, it seems likely that if our race survives long enough it will develop means of ultra-high-speed transport. Who knows how far our dreams will take us?

25

Is the Universe a Donut?

Someday spacecraft will be powerful enough perhaps to journey at extraordinary speeds, spanning the vast interstellar voids. Once space travel is swift enough our descendants may establish thriving extraterrestrial colonies, not just on nearby planets such as Mars, but also on worlds orbiting distant stars. At near-light velocities, relativistic time dilation would kick in, enabling voyagers to age much more slowly than they would on Earth and permitting them to survive flights that would otherwise be far too long. Hence, for spacecraft moving at fast enough speeds, the nearest stars, according to their calendars, would potentially be only a few months away. Perhaps our descendants will even learn how to master the fabric of space and time itself, constructing traversable wormholes that would permit phenomenal shortcuts from our region to another. Future technology could lay the groundwork for an extensive galactic civilization.

Throughout our galaxy, there are likely numerous habitable worlds ripe for exploration and countless other planets that could be rendered habitable through terraforming. Although astronomers have yet to identify planets with conditions similar to Earth's, over the past decade they have discovered hundreds of larger planets, similar in mass to Jupiter or Saturn. Only a handful of objects found so far are comparable in size to the smaller planets in our solar system, and these have much different orbital conditions from those on Earth. The reason astronomers haven't yet located less bulky objects with Earth-like conditions has more to do with the limitations of current techniques rather than their lack of existence. As planet hunting continues, numerous worlds similar to Earth are bound to turn up. A recent study by the researchers Sean Raymond of the University of Colorado at Boulder and Avi Mandell and Steinn Sigurdsson of Penn State University indicates that more than one-third of the systems with giant planets harbor Earth-like worlds as well,[1] potentially with breathable air and drinkable water. But do they have edible cream-filled crullers with chocolate sprinkles, served with frosted mugs of root beer with just a hint of foam on top? Alas, there are some questions science still can't answer.

NASA's Space Interferometry Mission (SIM) PlanetQuest program, due to be launched in the 2010s, is designed to scope out the closest stars for planets similar in mass to Earth, with orbital distances permitting moderate temperatures. It will target familiar stars such as Sirius and Alpha Centauri, hoping to find signs of worlds with just the right girth and range of climates. Could the Dog Star ironically be twirling around with a planet full of people (or the equivalent) on its gravitational leash? Or, if not, could there be a world full of water and vittles that would offer attractive fare for any Homeric voyagers on future space odysseys?

Once the Milky Way is explored, perhaps ships will tackle the even more formidable expanses between galaxies. Ultimately, human civilization, if it isn't challenged by other beings or destroyed through its own foolishness, could spread out throughout the cosmos and test the limits of space itself (if there are any). Our technology might develop until we become a vast, powerful intergalactic society, capable of resolving the deepest quandaries ever known. Only then could we definitely answer what is perhaps the ultimate question: "Is the universe shaped like a donut?"

This last question pertains to an idea attributed to Homer and mentioned by guest star Stephen Hawking in the episode "They Saved Lisa's Brain." In the episode, Lisa joins Springfield's chapter of the brainy organization Mensa, which also includes Professor Frink, Principal Skinner, Comic Book Guy, and businesswoman Lindsay Neagle. When Mayor Quimby temporarily abdicates his office due to a scandal, the Mensans take over, consistent with a city charter requirement that a learned council of citizens assume mayoral duties in the event of his absence. They vow to remake Springfield into a perfect society. The prospect of experiencing a blossoming utopia attracts the attention of Stephen Hawking, who (in his first animated appearance on the show) decides to visit and see it for himself.

As executive producer, Al Jean explained the decision to invite Hawking on the show: "We were looking for someone much smarter than all the Mensa members, and so we naturally thought of him. He seemed pretty interested in coming on right away."[2]

Hawking arrives just in time to see the townspeople revolting against new ordinances suggested by Frink and Comic Book Guy, such as banning many sports and restricting mating to once every seven years like *Star Trek*'s Vulcans do. Hawking escapes from the mayhem, rescuing Lisa in the process, using a flying device attached to his wheelchair. Upon Marge's suggestion, he and Homer segue over to Moe's tavern to relax and converse. Later Hawking is seen telling Homer, "Your theory of a donut-shaped universe is intriguing. . . . I may have to steal it."

In mathematics, a donut shape is known as a torus, the three-dimensional generalization of a ring. A ring lies in a single plane, so topologically speaking there is one closed path around it that lies just outside it (a loop around the ring). Because a torus has one more dimension, you can travel along closed paths around it in two perpendicular directions. If you imagine a donut on a plate, one of these is a larger loop around the periphery, parallel to the plate, and the other is a smaller loop through the hole, toward and away from the plate. The generalization of a torus, any closed curve spun in a circle around an axis, is called a *toroid*. Curiously, there are genuine scientific theories that the universe is toroidal.

Modern cosmology, the science of the universe, is mathematically modeled through solutions to Einstein's general theory of relativity. Recall that general relativity explains gravity through a mechanism in which matter curves the fabric of space and time. It is expressed in terms of an equation that relates the geometry of a region to its distribution of mass and energy. For example, an enormous star warps space-time much more, and therefore bends the paths of objects in its neighborhood by a greater amount, than does a tiny satellite.

Soon after general relativity was published, a number of theorists, including Einstein himself, delved for solutions that could describe the universe in general, not just the stars and other objects within it. The researchers discovered a plethora of diverse geometries and behaviors, each a distinct way of characterizing the cosmos. Some of these models imagined space as resembling an unbounded plain, something like the flat landscapes of Kansas and Nebraska, only uniform in three directions, not just two. Two parallel straight lines, in such a spatial vista, would just keep going in the same direction indefinitely, like prairie railroad tracks. Physicists call these flat cosmologies.

Other solutions possess spaces that curve in a saddle shape, technically known as hyperbolic geometries with negative curvature. This curvature couldn't be seen directly, unless you could somehow step out of three-dimensional space itself, but rather would make itself known through the behavior of parallel lines and triangles. In a flat geometry (called Euclidean), if you draw a straight line and a point not on it, you can construct just one single line through that point parallel to the first line. For a saddle-shaped geometry, in contrast, there are an infinite number of parallel lines fanning out from that point, like the tracks out of Grand Central Station. Moreover, while triangles in flat space have angles that add up to 180 degrees, in saddle-space the angles add up to less than 180 degrees.

Yet another possibility, called positive curvature, resembles the spherical surface of an orange. Like the saddle-shape, its form could be seen only indirectly, through altered laws of geometry. In general, curved geometries are called non-Euclidean because they do not follow all of the Greek mathematician Euclid's geometric postulates. In the case of positive-curved spaces, there are no parallel lines, and triangles possess angular sums of more than 180 degrees.

To understand these differences, slice an orange in half widthwise and cut the top half into quarters. Pick up one of the slices and look at the skin, and you'll notice that it is bounded by two edges that start out seemingly straight and parallel (where the widthwise slice was made) but end up meeting at the top. They are like any two lines of longitude on Earth, appearing parallel near the equator, but converging at the North Pole. This demonstration shows that no two lines on a positively curved surface are truly parallel.

What about lines of latitude, or the equivalent produced by slicing an onion in repeated widthwise segments? They appear parallel enough, never meeting. Strangely enough, on the spherical surface of Earth, they aren't true lines because they do not comprise the shortest distance between two points, technically known as geodesics. If you want to experience this yourself, purchase a ticket on a nonstop flight from Vancouver to Paris, both approximately the same latitude. Chances are, the flight will veer north, then south, rather than maintaining close latitude, because minimizing distance requires taking an orange-slice path—a geodesic—rather than an onion-slice path. It is these geodesics that must always meet somewhere, as seen in flight pattern maps.

Because geometry, in general relativity, influences dynamics, the shape of the cosmos bears greatly on its destiny. The vast majority of astronomers believe that the universe, regardless of its shape, started off as an ultra-dense point of extremely compact, perhaps infinitesimal, size, called the big bang, and expanded to its enormous present-day size. The exact manner of this expansion, and where it will ultimately lead, is partly determined by what geometry the universe possesses. If spatial geometry were the only determinant, then by knowing if the universe has negative, zero, or positive curvature you could predict if it will expand forever (in the case of negative or zero curvature) or someday reverse its expansion and contract back down to a point (in the case of positive curvature).

Geometry, however, cannot be the only influence on the dynamics of the universe. Another factor is an antigravity term, the cosmological constant, which was first suggested by Einstein. This term has come into prominence in recent years with the discovery by Adam Riess, Saul Perlmutter, Brian Schmidt, and their coworkers in various research teams that the universe is not just expanding, but is also currently speeding up in its expansion. This cosmic acceleration cannot be explained through geometry, but requires an additional outward boost, represented by the cosmological constant and known as dark energy. Models with a cosmological constant can have zero, negative, or positive curvatures, with the specific geometry affecting how and when the influence of the dark energy dominates the dynamics.

You might wonder why in this discussion we have mentioned flat shapes, saddle shapes, and orange shapes, but not yet donut shapes. It turns out that there has been traditionally much greater interest in hyperplanes (generalizations of infinite, flat surfaces), hyperboloids (generalizations of saddle shapes), and hyperspheres (generalizations of orange shapes) than in toroidal, donut-shaped cosmologies. Why are orangelike shapes, for instance, more favored in the literature than donuts? Looking at donut ingredients, some might think that this is a blatant example of the anti-trans-fatty-acid movement sweeping many locales, coupled with a bias toward the ascorbic acid (vitamin C) found in oranges. Surely it would be unwise to discriminate against models of the universe simply because of their passing resemblance to certain deep-fried pastries.

Actually, the bias in favor of hyperplanes, hyperboloids, and hyperspheres has more to do with their mathematical simplicity than anything else. They represent the most basic isotropic (appearing the same in all directions) three-dimensional surfaces, possessing the simplest topologies. Topology is different from geometry in that it concerns itself with how surfaces are connected, not their specific shapes and sizes. For example, topologically speaking, solid footballs, baseballs, basketballs, and even books about sports are all equivalent because they don't have holes through them, and you could theoretically transform one into another (assuming they were elastic enough) without cutting. Donuts, coffee cups with handles, tires, and hollow frames each have single holes, and therefore share common topologies distinct from continuous objects. Even if they are stretched the holes are still there.

A flat two-dimensional planar geometry—a square, let's say—can be transformed into a cylinder by identifying the far left side with the far right side, essentially gluing the two sides together. If an object travels far enough to the left, it ends up on the right. Something moving continuously to the left or right would experience the same region again and again in periodic fashion, like the animation loops common to cartoons from the 1960s and 1970s. Used to save time and effort, animation loops occur when the characters pass by the same background scenes again and again. For example, when Fred and Barney from *The Flintstones* drove down a road, they seemed to encounter the same array of rocks and trees over and over again. If you could explore a cylindrical universe, surviving somehow for tens of billions of years while traveling in what appeared to be a straight line, you'd have the same repetitive experience. Although you'd imagine that you're plowing directly ahead, you'd eventually circumnavigate space and pass the same array of galaxies once more in a topological déjà vu.

Space could be even more interconnected than that. Take a vertical cylinder and connect its upper and lower circles; what you get then is a torus. Now there are two perpendicular ways you can loop around the space: left-right and up-down. It's a bit like the 1980s arcade mainstay, the game of Pac-Man and its variants. When the colorful moving blobs exit the maze through any border portal, they miraculously pop up on the other side. Show them the back door and they gleefully return through the front, begging for more quarters.

An even more intricate arrangement links the extremes of all three spatial dimensions into a kind of "über-donut." Imagine space as a colossal cube; these connections would equate to the left and right, top and bottom, and front and back faces. Such a layout, a generalization of the torus with a three-dimensional instead of a two-dimensional surface, would be hard to visualize. Paradoxically, it merges a straightforward "flat" geometry (in the sense that parallel straight lines remain straight and parallel) with a mind-bogglingly complex topology.

Picture living in a house in which an ascending stairway in your attic leads to your basement, your front window has a scenic view of your rear kitchen, and your next-door neighbors are yourself. If the pipes under your living room happen to leak, the water would trickle down through all the lower levels, return through the upper floors, and ruin your living room furniture. Because there'd be nothing coming in from the outside world, everything in your residence would need to be recycled. You'd never be able to leave, just make the rounds through its doors and rooms again and again. Such would be life in a toroidal abode—not recommended for the claustrophobic. (Robert Heinlein wonderfully describes a similar situation in his story "And He Built a Crooked House.")

Could the entire universe have such a topology? The most reliable current data on the shape and configuration of space stem from missions to measure the cosmic microwave background (CMB), the cooled-down relic radiation from the big bang. The universe began its life very small, very hot, and very mixed up. Particles of matter and energy were bound together in a sizzling gumbo. Then, approximately 380,000 years after the initial burst, the stew cooled enough for complete atoms (mostly hydrogen) to coagulate, leaving the leftover photons (particles of light) as a kind of broth. At the point of separation, known as recombination, the matter was in some places a bit lumpier than in others, making the energy broth slightly uneven in temperature. These minute temperature differences have persisted throughout the ages, while the expansion of the universe has cooled down the energy broth significantly. From thousands of degrees Kelvin (above absolute zero) it's been reduced to a mere 2.73. Now it's a frigid backdrop of radio waves distributed throughout the universe.

The CMB was first discovered in the mid-1960s by the Bell Labs researchers Arno Penzias and Robert Wilson. While completing a radio wave survey, their horn-shaped antenna picked up a strange hiss. After they reported the result to the physicist Robert Dicke of Princeton, he calculated its temperature and found that it matched the predictions of the big bang theory. This discovery confirmed the existence of an ultra-hot beginning to the universe. It was not precise enough, however, to reveal the fine details of the primordial distribution of matter and energy.

A far more detailed examination of the CMB came in the early 1990s, thanks to the Nobel prize–winning work of John Mather and George Smoot. Using NASA's Cosmic Background Explorer (COBE) satellite, Mather and his team of researchers mapped out the precise frequency distribution of the microwave background radiation and established, beyond a shadow of a doubt, that it matched precisely what would be expected for a once-fiery universe cooled down over billions of years. Smoot and his group discovered a mosaic of minute temperature fluctuations (called anisotropies) throughout the sky, pointing to subtle early differences in the densities of various regions of the cosmos. These fluctuations showed how in the nascent universe slightly denser "seeds" existed that would attract more and more mass and eventually grow into the hierarchical structures (stars, galaxies, clusters of galaxies, and so forth), that we observe today.

The quest to map out the ripples in the CMB with greater and greater precision has continued throughout the past two decades. Uniquely, these provide a wealth of accessible information about the state of the cosmos many billions of years ago. It's like a rare cuneiform tablet that, with improving translations, provides richer and richer insights into ancient history each time it's read.

In 2001, the Wilkinson Microwave Anisotropy Probe (WMAP) was launched, offering an extraordinarily detailed mapping out of the CMB. From these data, astronomers have assembled an ultra-refined snapshot of the matter and energy distribution of the early cosmos. This information has furnished critical resolution of many long-standing cosmological riddles. For example, in the decades before WMAP there was considerable disagreement as to the age of the universe since the big bang. WMAP pinned down the value to be approximately 13.7 billion years—a fantastic achievement in the history of scientific measurement.

What of the shape of space? WMAP says much about that, too. Astronomers have gleaned the specific geometry of the universe by examining how the brightest patches in the CMB are stretched out or compressed in angle compared to what you would expect for pure flatness. While positive curvature would stretch these spots to 1.5 degrees and negative curvature would compress them to 0.5 degrees, zero curvature (flat) leaves them at 1 degree across. The third case appears to be true, so, based on that litmus test, space seems indeed to be flat.

In 1993, the U.C. Berkeley researchers Daniel Stevens, Donald Scott, and Joseph Silk proposed a way of sifting through CMB data to assess the topology of space as well. In their paper "Microwave Background Anisotropy in a Toroidal Universe," they showed how a universe with a multiply connected, toruslike topology would force the radiation into certain detectable wave patterns. Because such patterns seemed to be absent from the COBE data, the researchers did not find support for a toroidal cosmos.

Later work by Neil Cornish of Case Western, David Spergel of Princeton, and Glenn Starkman of the University of Maryland extended this technique to consider a wider range of possible topologies. Such a method has been applied to the WMAP results, examining the possibility that it could have a complex topology—not a toroid perhaps, but rather a dodecahedron (a bit like a soccer ball, but with all sides equivalent in size and shape). Although preliminary data (analyzed in 2003) seemed to rule out this model, more recent looks at the WMAP findings have revived the idea that if you venture far enough out into space you'll return to your starting point. Hence Homer's donut theory may have at least a sprinkling of truth: the universe could indeed have loops.

If the universe is truly loopy, what does it loop around? The two-dimensional surface of a sphere curves along a third dimension. Hence, fruits have cores as well as skins. What, then, would lie at the core of a looped three-dimensional cosmos? Could there be a higher spatial dimension beyond the limits of observation?

26

The Third Dimension of Homer

Homer's most grandiose schemes often fall flat. Despite efforts at self-improvement, many accuse him of lacking depth. When he flees from danger, it almost seems like he's just a shadow of a man. His foolish antics are cartoonish. You could accurately say that, despite his girth, he's one dimension short of being fully fleshed out.

Let's put these character limitations in perspective. These traits are not really his fault; he's just drawn that way. It's not what he does; it's the two-dimensional space in which he does it. The state of cartooning at the time *The Simpsons* was created did not allow for a weekly animated series with characters appearing realistically to be three-dimensional. Nor have sophisticated television cartoons moved in that direction today.

When the show began in the late 1980s, 3-D computer animation was barely in its infancy and was notoriously expensive. By that point only a few films had used computer-generated imagery (CGI), most famously a 1982 fantasy movie called *Tron* that cost $20 million to make and flopped at the box office. An epic about a programmer who is sucked into his own computer and becomes immersed in a jumble of otherworldly geometric images, *Tron* pioneered the idea of exploring virtual realms in a feature film. Nevertheless, only about fifteen minutes of what moviegoers watched constituted purely computer-generated sequences; the rest of the film featured more traditional special effects.

The failure of *Tron* to woo critics and attract large enough audiences to earn back its substantial costs scared off major studios from developing CGI features for quite some time. Gradually, 3-D computer animation techniques lowered in cost and improved in quality enough for Hollywood to invest in them again. Hence the abundance and popularity of such features today.

Whenever technology advances, the writers behind *The Simpsons* strive to keep up, aiming to achieve complete parody. In this case, the complete parody they achieved—the Treehouse of Horror VI segment called "Homer^3," which satirized both *Tron* and the "Little Girl Lost" *Twilight Zone* episode—was utterly brilliant. By transporting Homer from his flat, traditionally animated world into a computer-simulated three-dimensional realm, it reminded us that our own world could have unseen dimensions beyond our grasp.

The art historian Linda Dalrymple Henderson described the significance of this transformation: "Homer's transition from two to three dimensions as he walks through a wall provides a dramatic demonstration of the power of linear perspective and chiaroscuro modeling with light and shade, the two central artistic developments of the Italian Renaissance. At the same time it opens the door to discussions of our relationship to a higher, four-dimensional space, making clear the liberating potential of augmented dimensions."[1]

Stepping beyond the confines of ordinary space and into a higher dimension is a long-standing fantasy that dates back to the mathematical breakthroughs of the nineteenth century. The British mathematicians Arthur Cayley, James Sylvester, and William Clifford, and the German mathematicians Carl Gauss and Bernhard Riemann, among others, developed methods for extending three-dimensional structures into higher-dimensional entities. Geometries of more than three spatial dimensions have come to be known as hyperspace.

To help readers picture the concept of a higher dimension that is real but out of reach, in 1884 Edwin Abbott published *Flatland*, a novel about a two-dimensional world occupied by a society of geometric shapes. A. Square, the story's hero, lives and moves on a plane and does not realize that the universe extends beyond it. One day a sphere visits him to enlighten him about the third dimension. A. Square can't conceive of its existence until the sphere whisks him out of his plane and he witnesses it himself. Astonished by sights of both the insides and outsides of all the people, places, and things in his community, he returns and tries to convince others, only to have them think him mad. The lesson learned is that our inability to perceive a dimension doesn't exclude its actuality.

With the spookiness that characterizes *The Twilight Zone* in general, "Little Girl Lost," first broadcast in 1962, presents a more frightening excursion into a higher dimension. A father is dismayed to discover that his young daughter has vanished from her bedroom. Her voice sounds like it's coming from under the bed, yet she's not there. The family dog runs in after her and similarly disappears. Consulting with a physicist, the father learns that his daughter and his dog have somehow fallen through a portal into the fourth dimension. With his legs held carefully by the physicist, he dives into the portal and encounters a jumble of bizarre sights and sounds. Miraculously, in that transdimensional funhouse, he manages to locate his daughter and the dog and grab them. Quickly, the physicist pulls them all out of the portal, just in the nick of time before it closes forever.

"Homer^3" involves an analogous portal, only from the flatter-looking, less precisely sketched domain of traditional cartoons into the enhanced perspective and precision of computer graphics along the lines of *Tron*. Homer finds this portal behind a bookcase in his house and jumps in when faced with the horror of having to deal with his sisters-in-law. The hole reminds Homer, as he puts it, of "something out of that twilighty show." As soon as he enters, he acquires an extra measure of depth, which the artists depict by rendering his character with shading, perspective, and other three-dimensional visual cues. Surrounding him are solid geometric shapes and assorted equations from math and physics. Homer is utterly amazed—delighted and petrified at the same time—by the strange landscape and all the visual changes. Meanwhile, his family hears his disembodied voice emanating from various parts of the house but providing no indication as to where he actually is.

Much is made in the episode about the mathematical differences between Springfield and the computer-generated imagery. When Professor Frink tries to explain that Homer has slipped into the "third dimension," no one has any inkling as to which direction he is referring. Frink elicits gasps of astonishment when demonstrating how a cube generalizes the four-sided square into a six-faced object. It's as if they really are living in Flatland and are oblivious to the concept of space.

In actuality, almost all cartoons, traditional as well as computer-generated, try to simulate three dimensions, in some fashion, on a flat screen. ("Traditional" is a bit of a misnomer, because even "traditional animation" today makes use of computers during certain stages, meaning that the differences between it and CGI have narrowed.) If Springfield wasn't supposed to be in some ways three-dimensional, the opening sequence of the show, with clouds parting up in the sky, Bart writing on a vertical blackboard seen through a vertical window some distance away, Homer tossing a fuel rod out his car window, and the family converging on their driveway from all different directions would make absolutely no sense. They certainly couldn't fly on airplanes and spaceships, high above the Earth's surface, as shown in some of the episodes. Instead, they'd be navigating around a planar maze like Pac-Man. Perhaps Homer would blink now and then, change color to blue, and gobble up a flat donut or two, but otherwise a version of the show meant to look like it was completely two-dimensional would be quite dull. Luckily, the show maintains the illusion of three dimensions through layout, camera angles (scenes looking as if they are viewed through different vantage points), and a measure of shadows, shading, and perspective. So the leap Homer makes does not truly increase his dimensionality, rather just the way it is depicted.

After spending some time in the *Tron*-like geometric world and accidentally engendering a black hole, Homer starts to panic. Tying a rope around his waist, Bart rushes through the portal to help. Only the black hole's gaping mouth separates the two. But, alas, when Homer tries to leap across the abyss, the gulf is too wide and he plunges toward his doom. Surprisingly enough, Homer's doom turns out to be a mundane city street in the "real world" (using actual street footage). Mundane except for its erotic bakery, that is, where the "real Homer" (a costumed actor) ends up, bringing closure to this shapely episode.

Could objects really travel through hidden portals into a higher dimension? What once seemed pure mathematical abstraction or even mysticism has now come into its own as a legitimate scientific question. String and membrane theories, developed as unified visions of nature by physicists such as John Schwarz of Caltech, Michael Green of Queen Mary College, and Ed Witten of Princeton (along with others too numerous to mention), envision minuscule energy vibrations of various frequencies as the building blocks of all things. To encompass the four known fundamental natural interactions—gravity, electromagnetism, and the weak and strong forces, and for other technical reasons—these tiny strings must oscillate in a world of ten or eleven dimensions. Three of these dimensions represent the traditional modes of movement in space, and the fourth is time. These are the four physical dimensions the scientific community, even non–string theorists, generally accepts. String theorists suggest an additional six dimensions that are curled up so tightly they could never be directly observed. Like viewing a hairball from Snowball from a miles-high vantage in Nepal, you simply couldn't discern the tightly wrapped strands. Hence, these tiny compact dimensions would not contradict clear evidence that space has but three perpendicular ways to move.

In addition to the ten dimensions string theory requires to be physically realistic and encompass the natural forces, recent versions of the theory have made room for at least one "large extra dimension"—large enough, that is, to measure in the laboratory. This extra dimension emerges in a way of combining various types of string theory (along with membranes) into a unified vision called M-theory. M-theory includes both extensive and rolled-up dimensions. Why is it called such? According to Witten, "M" stands for magic, mystery, or matrix. Roll up (or not); it's the magical mystery theory!

In 1998, the Stanford theoretical physicists Nima Arkani-Hamed, Savas Dimopoulos, and Gia Dvali suggested that such a non-curled-up dimension of approximately one millimeter in size could resolve a long-standing mystery in physics: why gravity is so weak compared to the other natural forces. The idea that gravity is much feebler than other forces such as electromagnetism may seem strange until you realize that the entirety of Earth's gravitational pull can't stop an iron thumbtack from being lifted into the air by a small household magnet.

The Stanford researchers' theory envisions that the observable universe is located on a membrane floating in space, often called a brane for short. A second brane lies parallel to the first, separated from the other by a millimeter-wide gap, called the bulk, that extends along a higher dimension. Familiar matter—made of what are called open strings—clings to the first brane and cannot pass through the bulk. Particles conveying the electromagnetic force and all other interactions except gravity are similarly trapped. Gravitons, on the other hand, the carriers of gravitational force, are composed of closed strings and are thereby enabled to travel through the bulk. Because gravity leaks into the higher dimension, its strength is vastly diluted compared to the other forces that don't leak. That explains the discrepancy in strength between gravity and the other natural interactions.

A number of experiments designed to test the existence of large extra-dimensions have been performed, and many more are planned. Experiments led by Eric Adelberger of the University of Washington used a delicate device, called a torsion balance, to see if the law of gravity deviated from its standard Newtonian form (the gravitational attraction between two objects is inversely proportional to their separation distance squared). Adelberger has found no such discrepancy down to scales much lower than one millimeter, seeming to rule out at least the simplest version of the large extra-dimension theory.

Other experiments, both at the Fermilab particle accelerator in Illinois and the Large Hadron Collider (LHC), soon to be opened in Switzerland, are designed to look for "little gravitons lost": gravity particles escaping along higher-dimensional pathways. These projects involve smashing elementary particles together, examining the output of the collision, and seeing if any decay products are absent that would have involved the production of a graviton. Just as a sudden drop-off in choke marks on Bart's neck could signal Homer's abrupt vanishing, the lack of certain characteristic footprints in particle decay profiles could be a sign of a graviton disappearing act.

If we do live on a brane floating in the void, and if other such branes exist, and assuming they are in any ways similar to our own, it could be possible that there are civilizations dwelling in these parallel universes. Then perhaps we could send modulated gravitational signals and attempt communication with these extra-branar worlds. Just as we use radio waves of varying amplitudes (signal strengths) and frequencies to relay messages through ordinary space, we could possibly create gravitational waves with various characteristics to transmit information through the bulk. Conceivably, we could even find ways of converting ourselves into modulated gravitational pulses and beam ourselves into a parallel reality. I'm sure at least some readers are saying at this point, "Whoa, like

I haven't tried that already, dude," and others are sending off for their licenses to become parallel realty agents. I wouldn't invest in transdimensional property just yet, however; the existence of branes is purely hypothetical.

Now it's time to bring this extremely speculative discussion to a close. Let's return to our own world of three spatial dimensions where there are scientific wonders enough for many lifetimes of exploration. Indeed, there are sufficient mysteries involving the mind, the body, and living things in general to stimulate anyone looking for curious questions to examine. Toss in the workings of machinery and the secrets of robotics. If that's not enough to think about, ponder the secrets revealed through astronomy. Physics, robots, life, and the universe—that's ample grist to chew on for even the heartiest intellectual appetites. Mmm, grist. Add clever animated stories illustrating these topics and more through the interactions of an odd but endearing family and townspeople no one could forget, and there you have the science behind *The Simpsons.*

Inconclusion: The Journey Continues

Using episodes of *The Simpsons* as stepping-stones, we've taken an amazing voyage from the fundaments of individual human lives to the components of our incredibly vast cosmos. By exploring issues in biology, physics, robotics, time, and astronomy, we've helped answer Moe's rhetorical question, "What's science ever done for us?" True, rhetorical questions aren't usually meant to be answered, but what the heck, we answered it anyway—on the house. Now Moe owes us a free pretzel stick.

Our scientific musings have given us ample grounds for cautious optimism about the future. Even if the human gene pool has flaws, perhaps our descendants will be lucky enough not to inherit these characteristics and will be spared, like Lisa, from genetic predestination. And if these failings cannot be avoided, possibly science will reach the point where our progeny could be replaced by look-alike androids. If these malfunction and destroy the Earth, perhaps the remaining humans will be able to escape, journey to other planets, and establish new colonies. Supposing these outposts are overrun by slobbering, tentacled aliens, we could elect them leaders and hope their spirits buckle under the weight of bureaucracy. Then if the technology exists, we could sneak off to other galaxies. If the aliens launch an accelerator beam, speeding up time, and the entire physical universe is doomed, with any luck we'll discover an interdimensional portal to a new reality. But what if that new reality doesn't have donuts or twenty-four-hour Kwik-E-Marts? Ah, there's the rub. That's why I counsel *cautious* optimism.

As I write these words, a new era (for the series at least) is about to begin with the launching of *The Simpsons Movie*. In a world of computer-generated animation, it's gloriously in 2-D, as one of its preview trailers pointed out. Given that the trailer featured rabbits, flowers, a rock, a hard place, and a pendulum-like wrecking ball, lovers of science are buoyant with hope—even ornithologists, for whom hope is the thing with feathers. Will the film maintain the same level of sophistication as the series while addressing scientific issues? And what of episodes in coming years and maybe even movie sequels? Given the delightful prospects for continuing to address science through *The Simpsons*, that's why this ending is inconclusive. Let's hope that the journey has only just begun.

ACKNOWLEDGMENTS

As a longtime Matt Groening fan, I'd like to thank him for his brilliant contributions to humor, from *Life in Hell* to *Futurama* and *The Simpsons*. It's incredible how his work, supplemented by the talents of so many great writers, artists, and actors, has remained so vibrant and funny after more than two decades (since the series began as a segment on *The Tracey Ullman Show*). The voice-acting of Dan Castellaneta, Julie Kavner, Nancy Cartwright, Yeardley Smith, Hank Azaria, Harry Shearer, and others on the show is truly amazing, bringing so many distinct personalities to life.

Thanks to my wonderful agent, Giles Anderson, and the outstanding editorial team at Wiley, including Eric Nelson, Constance Santisteban, and Lisa Burstiner, for their help and vision for this project. The faculty and administration of the University of the Sciences in Philadelphia, including Philip Gerbino, Barbara Byrne, Reynold Verret, and Elia Eschenazi, have been great supporters of my research and writing. I am most grateful to Daniel Marenda and Alison Mostrom for reading over the biology chapters and making useful suggestions. Thanks also to Joe Wolfe of the University of New South Wales for his clever contributions and Linda Dalrymple Henderson for her useful remarks.

At my house we have our own *Simpsons* fan club. Chief among the members are my sons, Aden and Eli, who have been careful to look out for the science in the show. Whenever anyone in the house mentions that *The Simpsons* is on, there is an utter stampede to the television set. It's a bit like the couch scene opening each episode, come to think of it.

I appreciate the support of other family members and friends, including my parents, Stan and Bunny, and my in-laws, Joe and Arlene. Above all, I'd like to thank my wife, Felicia, for her helpful insights, love, and support.

THE SIMPSONS MOVIE HANDY SCIENCE CHECKLIST

The Simpsons Movie offers an ideal opportunity to practice what we've learned about the science in the series. For those who plan to grab their portable astrolabes, pack their pockets with forbidden goodies from their local Kwik-E-Marts, purchase a coveted ticket, and slip surreptitiously into a showing (or for those of you who are watching it on DVD, cable, a fingernail implant, or some other weird format of the future), I've prepared this handy science checklist for your viewing pleasure. Note that as of this writing, the movie has yet to be released, so these questions are necessarily fairly general. Here are some of the science questions you might ask yourself during the showing:

1. Has there been a radioactive leak, core meltdown, or other type of environmental catastrophe? If so, explain what caused it and what could have been done to prevent it.

2. Do any of the animals in Springfield exhibit abnormalities? Could they be signs of mutation? If so, speculate on the cause of these mutations.

3. In trying to save his family from utter doom, does Homer demonstrate the ingenuity of an Einstein, the persistence of an Edison, the vision of a Darwin, or the quiet, understated heroism of an inanimate carbon rod?

4. Has Carl put his master's in nuclear physics to good use? What about Apu's computer science degree? Has Lisa received long overdue credit for her scientific know-how, such as an early Ph.D.? Are there other trained scientists in the movie, and are they working to their full potential?

5. Has Professor Frink put forth any fantastic new inventions? If so, explain the science behind these.

6. Does the level of evil expressed by Burns, Snake, Sideshow Bob, Fat Tony, or any of the other sinister characters reflect their nature or nurture?

7. Are there robots and/or aliens in the movie? Could they be said to be fully conscious and alert (like Homer, for example) or are they mindless automata?

8. Is time in the movie like an ever-flowing stream, gently carrying the characters from one scene to another, or more like a stagnant puddle full of algae with vicious red frogs hopping to and fro? Does the movie suggest that the past is doomed to repeat itself? Hint: See the movie at least a few times before arriving at a conclusion.

9. Is the person sitting in front of you in the movie theater wearing an annoying large hat with an odd pattern? If so, explain the manufacturing process behind it, and the psychology of taste in clothing. If not, could the person in front of you be refraining from wearing such a hat because of a desire to fit in? In that case, grapple with the sociology of conformity.

10. At the end of the movie, after the credits roll, are the characters in a state of suspended animation? Compare and contrast this condition with cryonic freezing, death, and attending seminars on real estate management.

SCIENTIFICALLY RELEVANT EPISODES DISCUSSED IN THIS BOOK (LISTED BY CHAPTER)

1. "Lisa the Simpson," season 9, written by Ned Goldreyer, directed by Susie Dietter.
2. "E-I-E-I-(Annoyed Grunt)," season 11, written by Ian Maxtonc-Graham, directed by Bob Anderson.
3. "Two Cars in Every Garage, Three Eyes on Every Fish," season 2, written by Sam Simon and John Swartzwelder, directed by Wesley Archer.
4. "The Springfield Files," season 8, written by Reid Harrison, directed by Steven Dean Moore.
5. "In the Belly of the Boss," Treehouse of Horror XV, season 16, written by Bill Odenkirk, directed by David Silverman.
6. "The Genesis Tub," Treehouse of Horror VII, season 8, written by Ken Keeler, Dan Greaney, and David X. Cohen, directed by Mike B. Anderson.
7. "Lisa the Skeptic," season 9, written by David X. Cohen, directed by Neil Affleck.
8. "The Wizard of Evergreen Terrace," season 10, written by John Swartzwelder, directed by Mark Kirkland.
9. "The PTA Disbands," season 6, written by Jennifer Crittenden, directed by Swinton O. Scott III.

10. "B.I.: Bartificial Intelligence," Treehouse of Horror XVI, season 17, written by Marc Wilmore, directed by David Silverman.
11. "I, (Annoyed Grunt)-bot," season 15, written by Dan Greaney and Allen Grazier, directed by Lauren MacMullen.
12. "Itchy and Scratchy Land," season 6, written by John Swartzwelder, directed by Wesley Archer.
13. "Fly vs. Fly," Treehouse of Horror VIII, season 9, written by Mike Scully, David X. Cohen, and Ned Goldreyer, directed by Mark Kirkland.
14. "Stop the World, I Want to Goof Off," Treehouse of Horror XIV, season 15, written by John Swartzwelder, directed by Steven Dean Moore.
15. "Time and Punishment," Treehouse of Horror V, season 6, written by David X. Cohen, Greg Daniels, Bob Kushell, and Dan McGrath, directed by Jim Reardon.
16. "Future-Drama," season 16, written by Matt Selman, directed by Mike B. Anderson.
17. "'Scuse Me while I Miss the Sky," season 14, written by Dan Greaney and Allen Grazier, directed by Steven Dean Moore.
18. "Don't Fear the Roofer," season 16, written by Kevin Curran, directed by Mark Kirkland.
19. "Bart vs. Australia," season 6, written by Bill Oakley and Josh Weinstein, directed by Wesley Archer.
20. "'Tis the Fifteen Season," season 15, written by Michael Price, directed by Steven Dean Moore.
21. "Bart's Comet," season 6, written by John Swartzwelder, directed by Bob Anderson.
22. "Deep Space Homer," season 5, written by David Mirkin, directed by Carlos Baeza.
23. "Life's a Glitch, Then You Die," Treehouse of Horror X, season 11, written by Donick Cary, Tim Long, and Ron Hauge, directed by Pete Michels.

24. "Hungry Are the Damned," Treehouse of Horror, season 2, written by Jay Kogen, Wallace Wolodarsky, John Swartzwelder, and Sam Simon, directed by Wesley Archer, Rich Moore, and David Silverman.
25. "They Saved Lisa's Brain," season 10, written by Matt Selman, directed by Pete Michels.
26. "Homer^3," Treehouse of Horror VI, season 7, written by David X. Cohen, John Swartzwelder, and Steve Tompkins, directed by Bob Anderson.

NOTES

Introduction

1. Robin McKie, "Master of the Universe," *Observer*, October 21, 2001, p. C1.
2. Sarah J. Greenwald, "A Futurama Math Conversation with Dr. Jeff Westbrook," *Appalachian State University Report*, August 25, 2005, www.mathsci.appstate.edu/~sjg/futurama/jeffwestbrookinterview.html, last accessed February 25, 2007.

3. Blinky, the Three-Eyed Fish

1. Tom Lehrer, "Pollution," in *Too Many Songs* by Tom Lehrer (New York: Pantheon, 1981).
2. "The Cities: The Price of Optimism," *Time*, August 1, 1969, p. 1.
3. *New York Times*, October 16, 1927, cited in E. W. Gudger, "A Three-Eyed Haddock, with Notes on Other Three-Eyed Fishes," *American Naturalist* 62 (no. 683, November–December, 1928), pp. 559–570.
4. Alexander Meek, "A Three-Eyed Dab [Hippoglossoides limandoides]." Report of the Scientific Investigations Northumberland Sea Fisheries Commission for 1909–1910, 1910, p. 44.
5. "Ruminations of a Codfish Forker," in *The Fishermen's Own Book*, vol. 8 (Gloucester, MA: Procter Brothers, 1928), p. 28. Cited in E. W. Gudger, "The Three-Eyed Haddock, Melanogrammus Aeglefinus, a Fake," *Annals and Magazine of Natural History*, vol. 6 (1930), p. 48.
6. Anne Marie Todd, "Prime-Time Rhetoric: The Environmental Subversion of the Simpsons," *Enviropop: Studies in Environmental Rhetoric and Popular Culture* (Westport, CT: Praeger, 2002), p. 72.

4. Burns's Radiant Glow

1. Roger M. Macklis, "The Great Radium Scandal," *Scientific American*, August 1993, p. 94.

5. We All Live in a Cell-Sized Submarine

1. National Academy of Sciences, Commission on Physical Sciences, Mathematics, and Applications, *Size Limits of Very Small Microorganisms: Proceedings of a Workshop* (Washington: National Academies Press, 1999).

6. Lisa's Recipe for Life

1. Jack Szostak, interview in Andrew Rimas, "His Goal: To Unravel the Origins of Life," *Boston Globe*, September 25, 2006, p. D1.

7. Look Homer-Ward, Angel

1. Pete Mason, "Stephen Jay Gould," *Socialism Today*, issue 67, July/August 2002, www.socialismtoday.org/67/gould.html, last accessed February 25, 2007.
2. William A. Dembski, "An Analysis of Homer Simpson and Stephen Jay Gould," Access Research Network, www.arn.org/docs/dembski1129.htm, last accessed February 25, 2007.
3. Charles Darwin, *On the Origin of the Species by Means of Natural Selection, or the Preservation of Favoured Races in the Struggle for Life* (London: John Murray, 1859), p. 287.

9. Perpetual Commotion

1. Mark O'Donnell, "The Laws of Cartoon Motion," *Esquire*, June 1980. Reprinted in Mark O'Donnell, *Elementary Education: An Easy Alternative to Actual Learning* (New York: Alfred A. Knopf, 1985).

10. Dude, I'm an Android

1. Cynthia Breazeal, interview in Shaoni Bhattacharya, "New Robot Face Smiles and Sneers," *New Scientist*, February 17, 2003, p. 20.
2. Alan M. Turing, "Computing Machinery and Intelligence," *Mind: A Quarterly Review of Psychology and Philosophy* 59, no. 236 (1950), p. 433.
3. Conversations between Ned Block, John Sundman, and "Jabberwacky" by Rollo Carpenter. Transcripts of the 2005 Loebner Prize Competition, http://loebner.net/Prizef/2005_Contest/Transcripts.html, last accessed February 25, 2007.

11. Rules for Robots

1. Isaac Asimov, *I, Robot* (Greenwich, CT: Fawcett Crest, 1950), p. 6.

13. Fly in the Ointment

1. Anton Zeilinger, interview in *Sign and Sight*, February 16, 2006, Lucy Powell and John Lambert, trans., originally in *Die Weltwoche*, January 3, 2006.

14. Clockstopping

1. Rick Strassman, *DMT: The Spirit Molecule* (Rochester, VT: Park Street Press, 2001), p. 234.

19. The Plunge Down Under

1. Joseph Wolfe, University of New South Wales, personal communication, September 5, 2006.

20. If Astrolabes Could Talk

1. Geoffrey Chaucer, *A Treatise on the Astrolabe*; Addressed to His Son Lowys (London: N. Trübner for the Chaucer Society, 1872), p. 1.

24. Foolish Earthlings

1. G. R. Shipman, "How to Talk to a Martian," *Astounding Science Fiction*, October 1953, p. 112.
2. Giuseppe Cocconi and Philip Morrison, "Searching for Interstellar Communications," *Nature* 184, no. 4690, September 19, 1959, pp. 844–846.

25. Is the Universe a Donut?

1. Sean N. Raymond, Avi M. Mandell, and Steinn Sigurdsson, "Exotic Earths: Forming Habitable Worlds with Giant Planet Migration," *Science* 313, no. 5792, September 8, 2006, pp. 1413–1416.
2. Al Jean, interview in Joshua Roebke, "Meet the Geeks," *Seed*, April–May 2006.

26. The Third Dimension of Homer

1. Linda Dalrymple Henderson, University of Texas, Austin, personal communication, December 31, 2006.

FURTHER INFORMATION

For more on the science behind *The Simpsons*, you may wish to visit the hallowed Simpsonian Institution in Washington, D.C., which some have nicknamed "The Nation's Attic's Back Closet." Unfortunately this collection is not yet located on the National Mall, but rather in the attic of a shopping mall, via the service elevator next to the Leftorium. You need to ask for Mr. Jeff Albertson and be buzzed in. Due to recent budget cuts it is open only on Leap Days, from 9 P.M. until 5 A.M. Bring a flashlight and a can of scorpion repellent.

If you can't make it to the museum, or don't even believe it exists, here are many useful books and Web sites relating to *The Simpsons* and scientific issues and their connections discussed in this book. Following is a wee sample of what's out there.

Books

Matt Groening, *The Simpsons: A Complete Guide to Our Favorite Family* (New York: HarperCollins, 1998). This definitive guide to the series, written by its creator, provides much useful information about the show and its characters. It has several informative sequels.

The following two books address philosophical and religious issues related to the series: William Irwin, Mark T. Conard, and Aeon Skoble, eds., *The Simpsons and Philosophy: The D'oh of Homer* (Chicago: Open Court, 2001), and Mark Pinsky, *The Gospel According to the Simpsons: The*

Spiritual Life of the World's Most Animated Family (Louisville, Kentucky: Westminster John Knox Press, 2001).

Web Sites

"The Simpsons," www.thesimpsons.com. This is the official site.

"The Simpsons Archive," www.snpp.com. The best archive of Simpsons material, from commentary to scripts, run by loyal fans.

"Math on the Simpsons," http://simpsonsmath.com. A splendid collection of references to mathematics on the show, run by Dr. Sarah J. Greenwald of Appalachian State University and Dr. Andrew Nestler of Santa Monica College.

"Joe Wolfe: Educational Pages," www.phys.unsw.edu.au/~jw/education.html. Not *Simpsons* related, but useful for understanding the physics behind "Bart vs. Australia" and other episodes.

INDEX